BENCHWORK

*This book is dedicated to my wife, Janice,
for her devotion in typing and checking
the manuscript for this text.*

Australia • Brazil • Japan • Korea • Mexico • Singapore • Spain • United Kingdom • United States

Benchwork
Robert G. Dixon

For product information and technology assistance, contact us at
Cengage Learning Customer & Sales Support, 1-800-354-9706

For permission to use material from this text or product,
submit all requests online at **cengage.com/permissions**
Further permissions questions can be emailed to
permissionrequest@cengage.com

Library of Congress Control Number: 80-66607

ISBN-13: 978-0-8273-1743-7

ISBN-10: 0-8273-1743-3

Delmar
5 Maxwell Drive
Clifton Park, NY 12065-2919
USA

Cengage Learning is a leading provider of customized learning solutions with office locations around the globe, including Singapore, the United Kingdom, Australia, Mexico, Brazil, and Japan. Locate your local office at: **international.cengage.com/region**

Cengage Learning products are represented in Canada by Nelson Education, Ltd.

For your course and learning solutions, visit **delmar.cengage.com**

Visit our corporate website at **cengage.com**

Printed in the United States of America
20 21 22 23 24   15 14 13 12 11
ED207

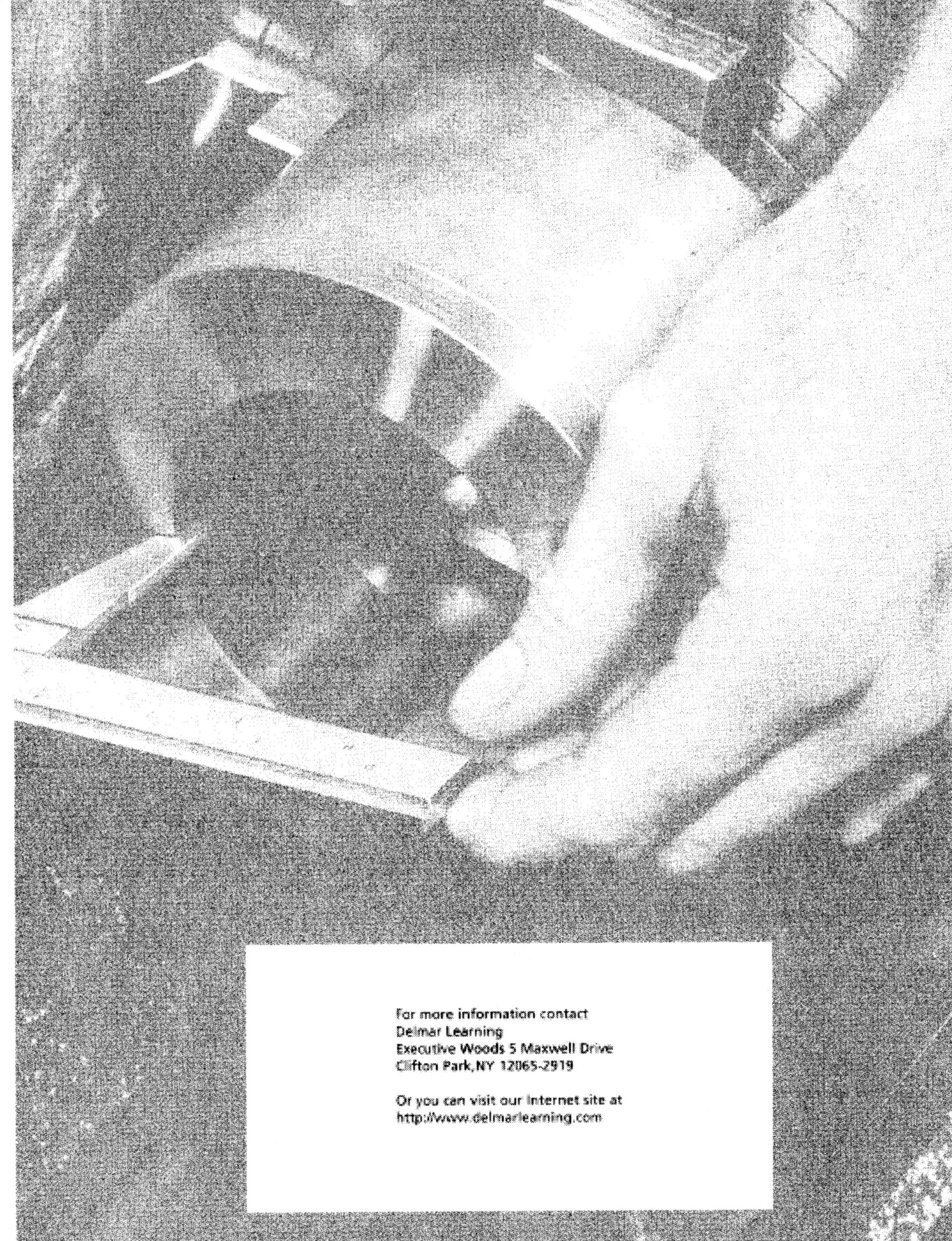

For more information contact
Delmar Learning
Executive Woods 5 Maxwell Drive
Clifton Park, NY 12065-2919

Or you can visit our Internet site at
http://www.delmarlearning.com

PREFACE

The benchworker is expected to make a profit when doing a job. If profits are to be made and productivity increased, the benchworker must learn to get the job done safely, accurately, and quickly. The benchworker should learn each bench procedure and tool in a manner that relates to profit.

BENCHWORK is written so that the student starts with the beginning essentials of benchwork practices and gradually moves to the more difficult benchwork processes. The textbook is designed to provide a basic understanding of the tools and procedures of applied benchwork related to the machine trades. Good work habits, attitudes, and positive human relations must also be learned and practiced while the tools, applications, and processes of benchwork are learned. BENCHWORK also provides the instructor with a complete, practical instructional tool for teaching benchwork.

Section one is an introduction to practical benchwork. Section two deals with the use and applications of measuring systems, measuring tools, and fitting principles. Section three describes various layout procedures and tools. The selection, care, and uses of bench hand tools (cutting and noncutting) are discussed in Section four. Section five discusses the power tools and special operations used in benchwork.

Learning Objectives are found at the beginning of each unit. These objectives tell the student and instructor what is to be learned after completing that unit. Photographs, drawings, and charts are included in each unit to aid the student in understanding the material that is presented. Also included throughout the text are step-by-step procedures to show how work is completed and how tools are properly used. Review Questions are found at the end of each unit and may be used to test the student's comprehension of the unit's content. Following the Review Questions, there are Activities which direct the student to apply what has been learned in that particular unit.

Safety precautions are highlighted throughout the text so the student is aware of the personal safety hazards when performing bench operations or using bench tools. Where applicable, metrics are included so that the student becomes familiar with the metric system of measuring.

For further study in the machine trades, the student may wish to refer to the following Delmar texts:

BASIC BLUEPRINT READING AND SKETCHING
BLUEPRINT READING FOR MACHINISTS — ADVANCED
BLUEPRINT READING FOR MACHINISTS — INTERMEDIATE
DRILLING TECHNOLOGY
DRILL PRESS WORK
ELEMENTARY BLUEPRINT READING FOR MACHINISTS
GRINDING TECHNOLOGY
INTERPRETING ENGINEERING DRAWINGS
INTERPRETING ENGINEERING DRAWINGS, METRIC
MILLING MACHINE WORK
TURNING TECHNOLOGY

ABOUT THE AUTHOR

Robert Dixon is a journeyman machinist and certified welder who has worked in the machine industry for 25 years. He began as an apprentice and worked through the shop to supervision. He has also owned his own company dealing with processing machinery, design, manufacture, sales, and repair. He has recently earned an A.S. degree in Industrial Technology.

For the past 6 years, he has been an Instructor in Machine Shop Technology at Chemeketa Community College in Salem, Oregon. He is an active member of the Oregon State Machinist Apprenticeship Council, the American Welding Society, the American Vocational Association, the Oregon Vocational Association, and the Oregon Vocational Trade-Technical Association. He is a senior member of the Society of Manufacturing Engineers and serves as faculty advisor for Chemeketa's Student Chapter. He was also honored as the Teacher of the Year by the Chemeketa Deaf Program for learning sign language to be able to communicate with deaf students in the machine shop program in 1977-78. He was voted the 1979-80 Industrial Educator of the Year by the Oregon Vocational Trade Technical Association and the 1980-81 Educator of the Year by the Oregon Vocational Association.

NOTICE TO THE READER

RIDGID

CONTENTS

Preface   iv

SECTION I   INTRODUCTION TO PRACTICAL BENCHWORK

Unit 1   Basics of Applied Benchwork   1
Unit 2   Materials   7
Unit 3   Material Identification and Storage   16
Unit 4   Mechanical Fasteners   23

SECTION II   MEASUREMENT SYSTEMS, TOOLS, AND FITTING

Unit 5   Units of Measure   37
Unit 6   Semiprecision Measuring Tools and Gages   42
Unit 7   Precision Measuring Tools and Gages   54
Unit 8   Tolerances and Fits   70

SECTION III   LAYOUT PRINCIPLES

Unit 9   Nonprecision and Semiprecision Layout Tools   79
Unit 10   Precision Layout Tools   91
Unit 11   Layout Procedures   97

SECTION IV   HAND TOOLS

Unit 12   Assembly Tools, Arbor and Shop Presses   105
Unit 13   Chisels, Files, Scrapers, and Abrasive Cloth   117
Unit 14   Drills, Reamers, Broaches, and Saws   129
Unit 15   Taps and Dies   144

SECTION V   POWER TOOLS

Unit 16   Power Hand Drills and Grinders   159
Unit 17   Power Saws   166
Unit 18   Bench and Pedestal Grinders   178
Unit 19   Special Benchwork Tools and Procedures   186

Appendix   194
Acknowledgments   203
Index   205

SECTION I
INTRODUCTION TO
PRACTICAL BENCHWORK

UNIT 1 BASICS OF APPLIED BENCHWORK

OBJECTIVES

After completing this unit, the student will be able to

- list the basic duties of a benchworker.
- list the tools used and the operations performed in benchwork.
- explain the proper care of bench tools.
- describe benchwork layout procedures with regard to accuracy and productivity.

INTRODUCTION

Benchwork is the start and finish of machine work. The machine trades can be divided into machine work and benchwork.

Machine work is work done in the shop by machine tools, such as lathes, milling machines, drill presses, saws, and grinders, Figure 1-1. *Benchwork* is the term used to describe work that can be placed on a bench or in a bench vise to be worked on, Figure 1-2. The operations done on this work mainly involve tools that are controlled by the hand. These tools are called *hand tools*. *Floor work* refers to larger-type work that cannot be placed easily on the bench. The same tools and processes as in benchwork are used, only on a larger scale.

DUTIES OF THE BENCHWORKER

1. Select the proper material indicated on the print.
2. Cut these materials to size.
3. Accurately lay out the workpiece for hand or machine operations.
4. Clean and finish the work.
5. Inspect the finished work for quality and accuracy requirements.
6. Assemble and properly fit parts together.
7. Test the completed workpiece assembly.

Attitudes and Safe Work Habits

Attitudes and safe work habits are two very important items in any place of work. Benchwork presents some challenging layout, fitting, and assembly tasks. Workers in the machine trades must think logically and apply good judgment to every task.

Most tools used in benchwork can be dangerous. If proper thought is not given before an act is made, serious damage or personal injury may result. The main cause of accidents in the shop is carelessness. The student should learn the accepted safe work habits at the same

Fig. 1-1 Machine work is often done on tools such as lathes.

Fig. 1-2 Benchwork is done with smaller hand tools, vises, and the bench shown.

time benchwork tools and methods are learned. Students are trained and expected to work in the following order:

1. Safely — so you and other workers may continue to work.
2. Accurately — to avoid costly mistakes.
3. Quickly — to make company profits.

Through practice, the student will become the skilled craftsperson needed by industry.

In most shops, the following work habits and attitudes are required:

- Maintain good attendance and be on time to work.
- Be willing and ready to work.
- Have concern and care for tools and equipment.
- Be courteous and show respect for employers and fellow workers.
- Be loyal and trustworthy.
- Work in a safe manner.

Specific practices dealing with the safe, efficient operation of individual shop tools are included in each unit of the text.

Safety Cautions are noted as each tool is presented.

Mistakes and Profits

When working in shops, mistakes must be avoided because:

- Materials are wasted.
- Shop time is lost two times.

   Loss 1 — first time the part was made wrong.

   Loss 2 — time to make the same part again correctly.

- Mistakes cost time, and time is money.

Profits must be made by a company to remain in business. From the profits come wages, benefits, new tools and equipment, and improved working conditions. The following may be used as a guide to profitable benchwork:

- Work to the degree of accuracy required on the print.
- Carefully plan the work.
- Carefully lay out the work.
- Check the layout carefully before starting work.
- Select the correct materials.
- Practice good, safe work habits.

BENCHWORK OPERATIONS AND TOOLS

The following operations are among the several procedures used in benchwork. These operations are explained in the units that follow.

Chipping	Filing	Laying out
Sawing	Fitting	Assembly
Drilling	Scraping	Disassembly
Reaming	Polishing	Testing
Tapping	Grinding	Inspection
Threading	Measuring	

The following are some of the many tools used in benchwork:

Hammers	Measuring tools
Vises	Assembly tools
Wrenches	Layout tools
Cutting tools	Inspection tools
Metal stamps	Screwdrivers
Small drills	Chisels
Hand reamers	Punches
Hand hacksaws	Hand taps
Files	Threading dies

The benchworker must be able to select the proper tools for a job. Skillful use of these tools is gained by knowledge, patience, and continued practical use.

GENERAL CARE OF BENCH TOOLS

Tools can perform many tasks beyond the power of our own hands. Below are suggestions for the proper care of bench tools.

- Place tools not being used in a safe place so they are not damaged or lost, Figure 1-3. Tools break and lose accuracy if they are allowed to drop on the floor or become dirty.

- Have a place in your tool box or shop for each tool, Figures 1-4 and 1-4A.

- Properly clean, adjust, and lubricate tools before putting them away. The tools will then be ready for the next use.

- Be careful with sharp tips and points on tools as they are easily damaged. Make sure these tips and points are kept in good

Fig. 1-3 Tool island for clean, safe tool storage

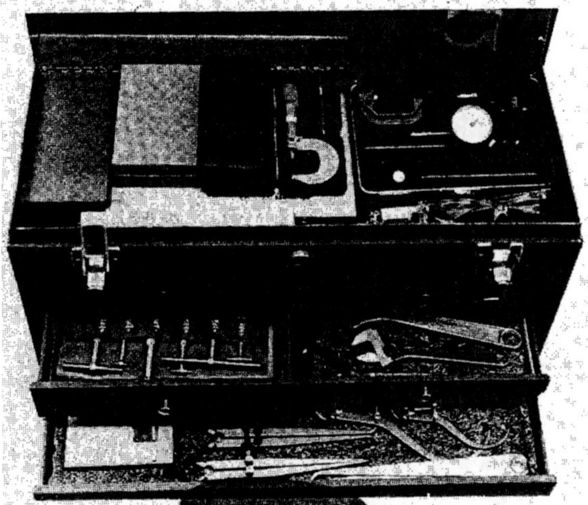

Fig. 1-4 Tools properly stored in a tool box

Fig. 1-4A Tools properly stored in a shop bench area

condition. Do not carry these tools in your pockets. This may cause personal injury.

- Keep electrical cords in good condition and away from possible damage.

PLANNING

A job must be properly planned before starting. One method is to make an outline or procedure to follow. Many companies have special forms called *operational sequences* or *process sheets*. These process sheets are rigidly followed. They assure job completion in a certain amount of time to the required accuracy.

Logical Steps to Complete A Job

A typical work plan would include the following:

- Material selections
- Tool selections
- Safety precautions
- Information such as speeds and feeds of machine tools and other special tooling or fixture setups.

At first, the work plan is lengthy. From practical experience, the outline becomes brief and at times, a mental process. The outline is a good work habit. It allows for safer work, fewer mistakes, productivity (work finished accurately in less time), increased profits, and personal satisfaction of a job well done.

BENCH REPAIR WORK

At times, benchwork involves repair of machines or parts. The following procedure may be used as a guide:

1. Diagnose the problem.
2. Make a sketch, take pictures of parts, or refer to a repair manual.
3. Disassemble the parts. Keep the parts in order.
4. Inspect the parts for wear.
5. Select new parts or repair worn and damaged parts.
6. Reassemble the parts and fit repaired parts as required.
7. Test the repaired unit or machine for the proper operation.

REVIEW QUESTIONS

A.   Multiple Choice

1. Work that is too large to be done on the bench is called:
   a. benchwork.
   b. machine work.
   c. floor work.
   d. table work.

2. How many times is shop time lost by an error which causes a scrapped workpiece?
   a. 1 time
   b. 2 times
   c. 3 times
   d. 4 times

3. The main cause of accidents in the shop is:
   a. loose clothing.
   b. carelessness.
   c. wearing gloves.
   d. a messy shop.

4. To work to the required degree of accuracy means to work:
   a. within print limits.
   b. slowly.
   c. very accurately.
   d. quickly.

5. What is the first procedure you should follow when repairing a machine or part?
    a. Make a sketch of the machine or part you wish to repair.
    b. Diagnose the problem.
    c. Refer to a repair manual.
    d. Inspect the part or machine for wear.

B. Short Answer

6. What is the best way to gain skill in using hand tools?

7. What happens to bench tools if they are allowed to drop or become dirty?

8. List six operations of benchwork.

9. List six tools used in benchwork.

10. List the three things a benchworker should do before putting tools away.

11. What is an *operational sequence*?

12. List the four items a typical work plan should contain.

13. Define the term *productivity*.

14. Name the first work habit a student must learn.

ACTIVITY

Review with your instructor the following operational-sequence outline to make the mount plate shown on page 6. This outline will help to get your workpiece completed more safely, accurately, and quickly.

1. Study the print carefully.

2. Select the proper kind and correct shape of material.

3. Cut material to correct size, length, and number of pieces.

4. Remove all sharp edges and burrs. Use a file with a handle or use an abrasive machine. Wear safety glasses.

5. Clean the layout surface free of all dirt, scale, and grease.

6. Apply a light coating of layout solution over the surface to be laid out.

7. Establish and mark accurate reference edges or baselines to work from. Edge A and Edge B are the reference edges. Mark the lower left-hand corner as your reference or base corner X.

8. Select necessary layout tools and organize them properly.

9. Measure, layout, and scribe the hole location centerlines.

10. Check your layout with the print. Correct any errors.

11. Prick punch the center points. Be careful to punch exactly at the cross of the vertical and horizontal centerlines. Wear safety glasses.

12. Scribe the reference circles.

13. Prick punch the circumferences of the scribed reference circles, especially on work which has a rough surface. The scribed circle may be removed by the rubbing of the cutting chips on the work surface during the drilling operation. However, the prick punch marks will remain as a reference check to accuracy.

14. Center punch the center points. This provides a larger indent for the twist drill to start cutting more easily.

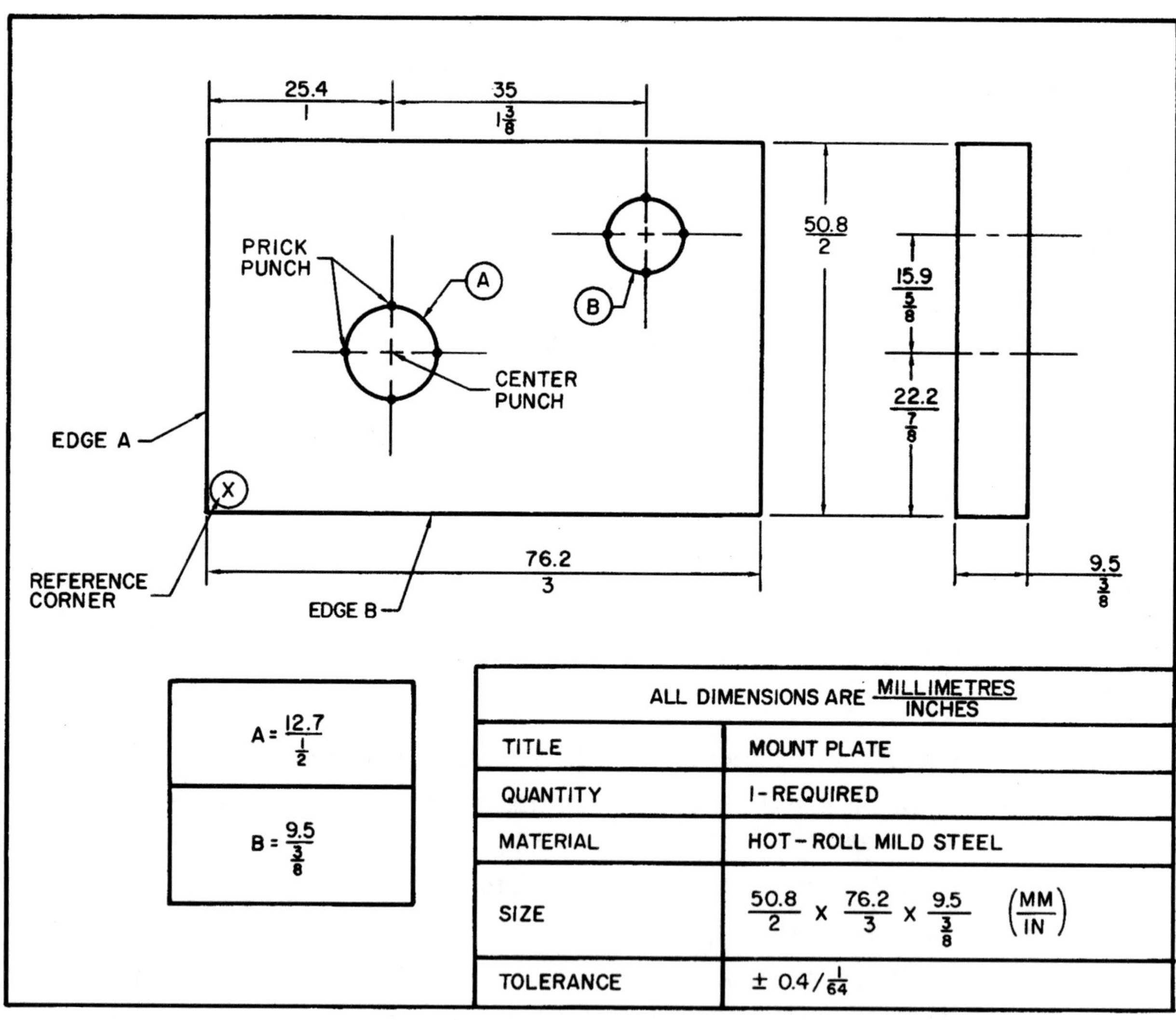

UNIT 2 MATERIALS

OBJECTIVES

After completing this unit, the student will be able to

- list the three basic forms of metal used in the shop.

- demonstrate how common material shapes are dimensioned and measured.

- identify the four general types into which metals are divided.

INTRODUCTION

Many kinds of materials are used in the shop to make the mechanical parts required for various machines and different types of tools. The kind of material to use depends on for what and where the part is to be used. Some parts are required to be made of a hard, tough material. Parts may be made of softer metals, plastics, or other nonmetallic materials. Some metals and materials can be surface treated with other materials to protect them from *corrosion* (rusting or weathering away). Parts may need to be made of a material to resist corrosion.

Because of the many uses of parts, it is extremely important to correctly identify and select the right kind of material called for on the print. This selection must be made before the job is started. Correct material selection saves labor, valuable shop time, and expensive materials.

BASIC FORMS OF METALS

The three basic forms of metals include bar stock, forgings, and castings.

Bar Stock

Bar stock is the most common form of shop metal and is made in many standard shapes and sizes, Figure 2-1. From these shapes various required parts are machined and assembled into machinery and tools. It is made in standard mill forms and includes sheet metal and other structural shapes. Each shape is dimensioned and measured as shown in Figure 2-2. Most suppliers have common shapes and sizes in stock. The steel supplier can furnish special lengths and sizes at an additional cost.

Pipe and tubing are two other metal shapes closely related to bar stock and often used in the shop. Both are available in ferrous, nonferrous, and alloy materials. Pipe

Fig. 2-1 Various shapes and sizes of bar stock

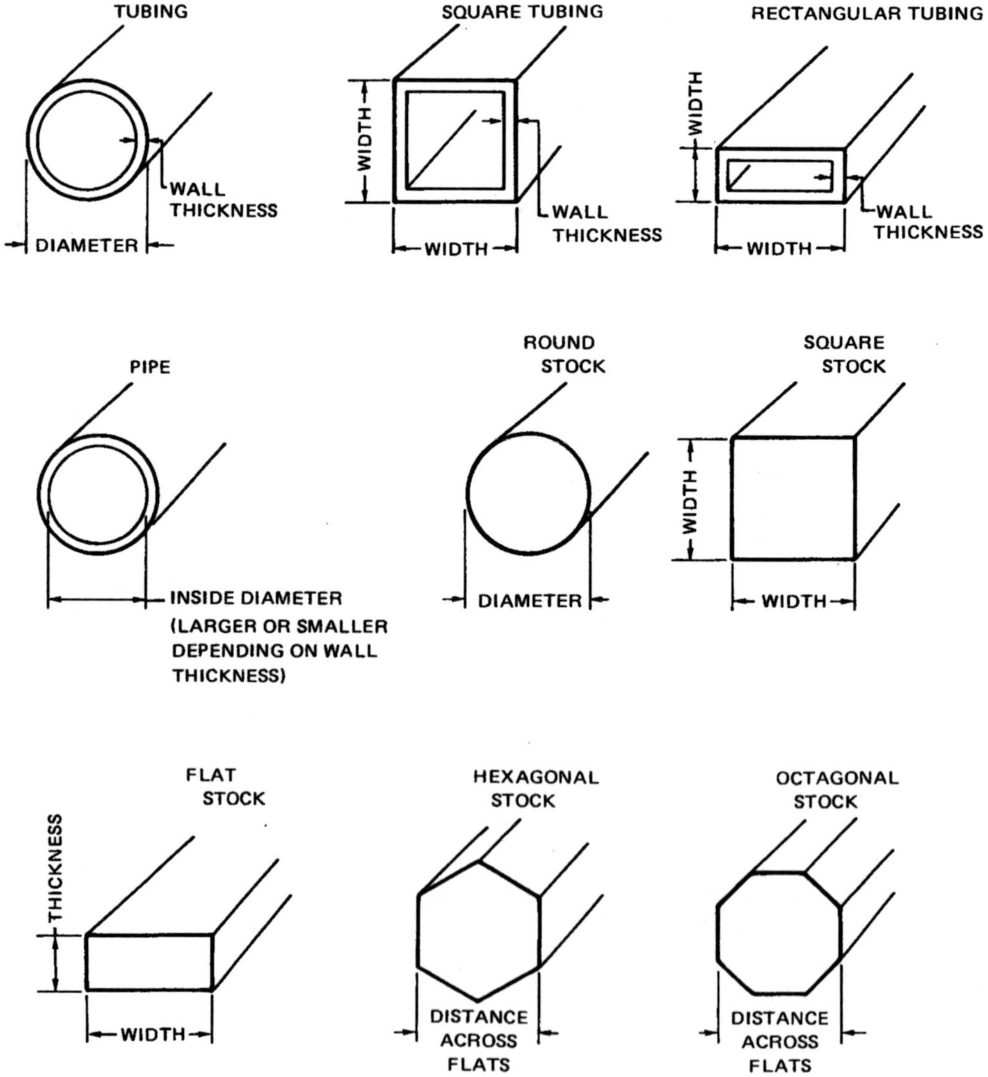

TUBING — WALL THICKNESS — DIAMETER

SQUARE TUBING — WIDTH — WIDTH — WALL THICKNESS

RECTANGULAR TUBING — WIDTH — WIDTH — WALL THICKNESS

PIPE — INSIDE DIAMETER (LARGER OR SMALLER DEPENDING ON WALL THICKNESS)

ROUND STOCK — DIAMETER

SQUARE STOCK — WIDTH — WIDTH

FLAT STOCK — THICKNESS — WIDTH

HEXAGONAL STOCK — DISTANCE ACROSS FLATS

OCTAGONAL STOCK — DISTANCE ACROSS FLATS

Fig. 2-2 Procedures to measure standard shapes of bar stock, pipe, and tubing

| STANDARD WEIGHT WROUGHT STEEL PIPE BLACK AND GALVANIZED | | | | | |
Size Inches	Diameter, External	Inches Internal	Thick-ness Inches	Wt., Lb., per Ft. Plain End	Threaded and Coupled
1/8	.405	.269	.068	.24	.24
1/4	.540	.364	.088	.42	.42
3/8	.675	.493	.091	.57	.57
1/2	.840	.622	.109	.85	.85
3/4	1.050	.824	.113	1.13	1.13
1	1.315	1.049	.133	1.68	1.68
1 1/4	1.660	1.380	.140	2.27	2.28
1 1/2	1.900	1.610	.145	2.72	2.73
2	2.375	2.067	.154	3.65	3.68
2 1/2	2.875	2.469	.203	5.79	5.82
3	3.500	3.068	.216	7.58	7.62
3 1/2	4.000	3.548	.226	9.11	9.20
4	4.500	4.026	.237	10.79	10.89
5	5.563	5.047	.258	14.62	14.81
6	6.625	6.065	.280	18.97	19.18
8	8.625	8.071	.277	24.70	25.55
8	8.625	7.981	.322	28.55	29.35
10	10.750	10.192	.279	31.20	32.75
10	10.750	10.136	.307	34.24	35.75
10	10.750	10.020	.365	40.48	41.85
12	12.750	12.090	.330	43.77	45.45
12	12.750	12.000	.375	49.56	51.15

Fig. 2-3 Dimensions of standard steel pipe (Standard Steel Company)

and tubing are also made in plastics and other nonmetallics.

Pipe is sized by its inside diameter and length. The size is usually larger or smaller than the nominal size, depending upon the wall thickness. The outside diameter must be kept a certain size. This is so threads cut on the pipe are a standard size, Figure 2-3.

Tubing is sized by its outside diameter, wall thickness, and length. The wall thickness is usually given in decimal parts of an inch. Tubing is also made in standard rectangular and square shapes of varying sizes and wall thicknesses, Figure 2-4.

Forgings

This method of metal form is used to produce parts which require great strength. Large, heavy presses force the steel into semifinished

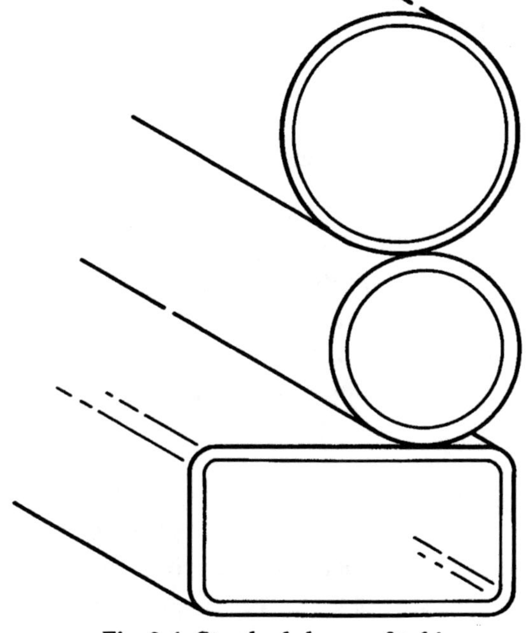

Fig. 2-4 Standard shapes of tubing

shapes. Little machine work is required to finish the forging to the required size, Figure 2-5.

Castings

Cast parts are made by pouring molten iron into molds of a product shape, Figure 2-6.

SPECIAL MATERIALS

There are many other special materials used in the shop other than bar stock, castings, tubing, and pipe. These benchwork materials include gaskets, lubricants, epoxies, shims, liquid sealants, metal fillers, fasteners, and many others, Figure 2-7. These materials are used in the assembly and repair of tools, machines, and component mechanisms. The print will name the material required in the bill of materials to complete the particular job. Suppliers provide complete catalogs for selection of needed materials. It is a good practice to set up and maintain a current library of these catalogs.

TYPES OF INDUSTRIAL METALS

Industrial metals are divided into four general types:

- Ferrous
- Nonferrous
- High temperature
- Rare and precious

FERROUS METALS

Ferrous metals are those which contain iron. Iron and steel are the more common ferrous materials used in the construction of tools and machinery.

Iron

Iron used in the shop is low cost and used in the form of cast parts where weight and mass are needed. Iron is mixed with varying amounts of carbon and alloy materials to make different kinds of cast iron.

Cast Iron. Cast irons have a rough, pitted surface and are dull gray in color. The outer surface, or scale, is hard and difficult to cut. The interior of cast iron cuts easily when the outer surface is removed. There are several types of cast iron available.

Gray cast iron is the least expensive type of cast iron and is used where cost is a factor. It is used to make large pipes, machinery frames, and large, heavy machine parts which require little impact strength. White cast iron

Fig. 2-5 Typical forged parts

Fig. 2-6 Typical cast part

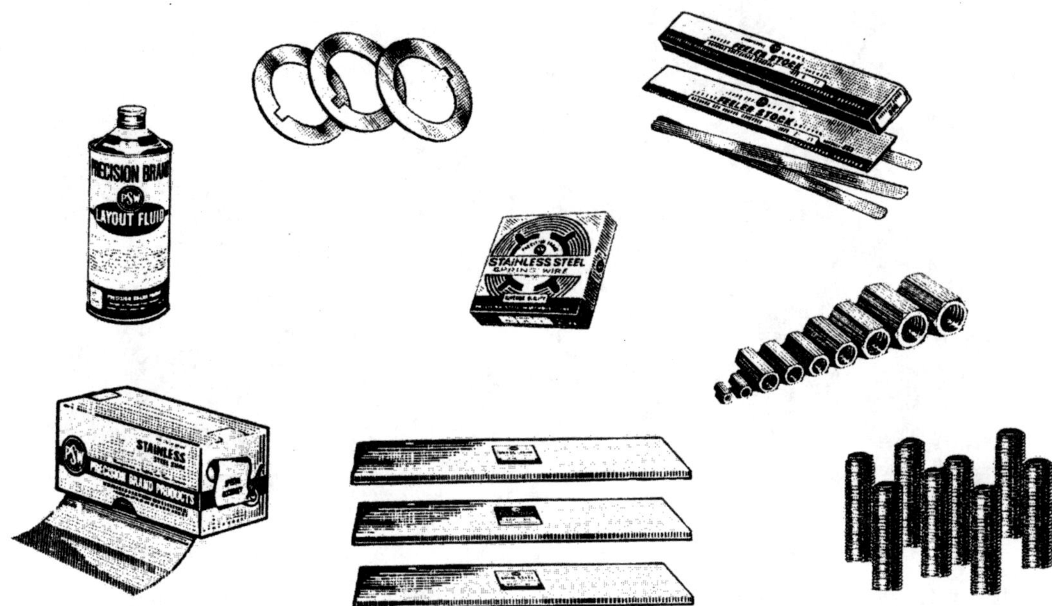

Fig. 2-7 Various products used in benchwork other than bar stock (Precision Brand Products)

is used where a surface must withstand abrasion and wear. Much white cast iron is used to make malleable iron.

Malleable iron is used where toughness is required in automobile and farm machinery parts. Ductile cast iron is similar to malleable iron. It is used to make rough castings for many kinds of machinery.

Wrought Iron. Wrought iron is a very tough and malleable metal with a very low carbon content. It bends easily, rusts slowly, and is easy to weld. The main uses are in wrought iron railings and outdoor furniture. Wrought iron is the purest form of iron in commercial use. It is often just called *iron*.

Steel

Steel is an alloy of iron and carbon with varying amounts of other elements. These elements are silicon, sulfur, phosphorus, and manganese.

Carbon steels are classified according to the amount of carbon they contain. Steels made with varying amounts of carbon are referred to as:

- Low-carbon steels — 0.04 percent carbon to 0.30 percent carbon

- Medium-carbon steels — 0.30 percent carbon to 0.60 percent carbon

- High-carbon steels — 0.60 percent carbon to 1.50 percent carbon

- Cast iron — 1.7 percent carbon to 4.5 percent carbon

Carbon is the element which gives steels, especially tool steels, the property which allows it to be hardened. Also, the carbon content largely determines the strength of a steel and its ability to hold a cutting edge. The carbon content normally varies from 0.50 percent to approximately 1.50 percent. This percentage measurement is termed *points*. Thus, 100 points equals 1 percent.

Low-Carbon Hot-Rolled Steel. Sometimes called machine or mild steel, low-carbon steel is one of the cheaper grades of steel. It contains a very low percent of carbon. This type of steel can easily be forged but does not machine smoothly. It is recognized by a

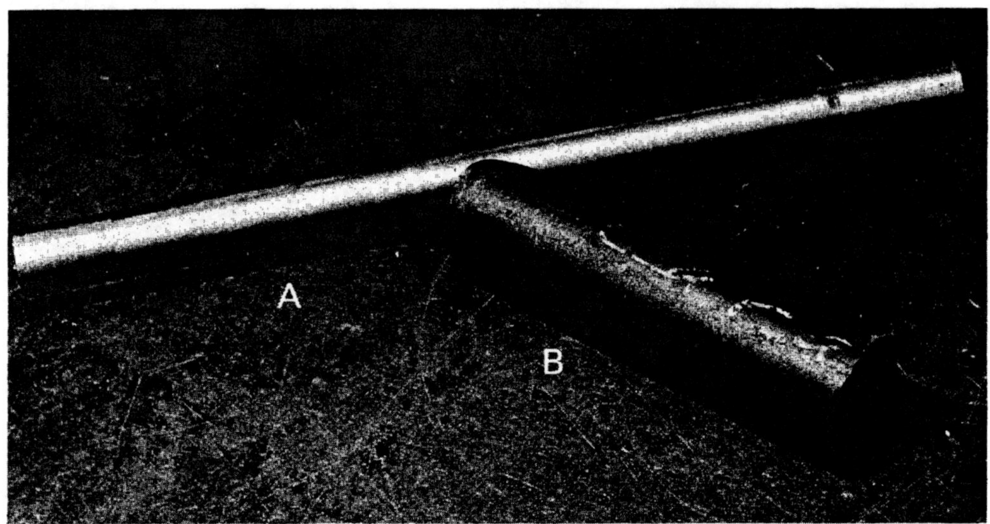

Fig. 2-8 (A) Cold-rolled mild steel, and (B) hot-rolled mild steel
(Note: Most steels are made available in either hot-rolled or cold-rolled condition.)

rough-oxide outer surface caused from hot rolling in the processing stage, Figure 2-8.

The diameter is usually left over *nominal size* (standard inch measurements of 1/64 (0.4 mm), 1/32 (0.8mm), 1/16 (1.5mm), 1/8 (3.1 mm), 1/4 (6.3mm), 1/2 (12.7mm), etc.). This is done so that when the oxide is machined off, a nominal size can be maintained. This material is used in parts which do not require much strength. It is also used for most bench metal and ornamental ironwork.

Cold-Rolled Mild Steel. This is another low-carbon steel and is processed by cold rolling, Figure 2-8. This process improves the outer metal surface by making it smoother and denser. This also improves the machinability of the steel. *Machinability* is the relative ease or difficulty involved in cutting steel.

The surface of cold-rolled mild steel is very close to nominal diameter sizes, usually a few thousandths of an inch under. This metal can only be used where cutting edges, hardness, or great strength are not required. The surface of this metal can be *case hardened* to a minimum depth by adding carbon under heat to the surface. The part would then need to be properly heat-treated. *Heat treating* is a heating and cooling process in which metal is made harder and tougher.

Medium-Carbon Steels. These steels are used for hammerheads, clamp parts, screwdrivers, sledges, crankshafts, and axles. This material can be heat-treated to obtain hard, tough surfaces.

High-Carbon Steels. These steels are used for making small tools or parts that need to be hardened and heat-treated (tempered). Tools which must hold a cutting edge, such as punches, chisels, taps, reamers, and twist drills, are made of high-carbon steel. High-carbon steel is also known as *tool steel* It is marked with manufacturer's labels, stamped or etched symbols, or color paint which identified the grade and kind of steel.

Alloy Steels. Alloy steels are steels that are made harder, tougher, or stronger than ordinary carbon steel by adding other metals to the steel. The basic alloying metals used are nickel, chromium, manganese, molybdenum, tungsten, and vanadium. Each alloy metal, when added in varying quantities, changes the properties of the steel.

A common alloy steel is known as *high-speed steel* (carbon from 0.70 percent to 1.5 percent, chromium, vanadium, molybdenum, tungsten, and cobalt). This self-hardening alloy steel is used to make cutting tools that have red hardness. *Red hardness* is the ability of a tool to retain edge hardness at high temperatures. Tungsten gives high-speed steel this quality. For example, a high-speed steel twist drill can be run faster than a high-carbon twist drill of the same size and retain its cutting edge longer.

The addition of the basic alloying metals will make changes in steel as follows:

- Nickel increases corrosion resistance.
- Chromium gives hardness, strength, wear, and corrosion resistance.
- Manganese increases hardness and toughness.
- Molybdenum increases elastic limit, impact strength, wear resistance, and fatigue strength.
- Cobalt causes the cutting edge to remain sharp by improving resistance to abrasion.
- Vanadium improves strength and toughness.

Stainless Steels. Stainless steels are made in more than 100 different alloys. Stainless steels are a straight chrome-iron alloy or a nickel-chrome alloy. The main characteristic of stainless steel is that it contains enough chromium to be corrosion resistant. Stainless steels do contain varying amounts of iron, therefore some grades can rust. Abrasion and wear resistance, along with hardness, strength, toughness and ductility are other desirable qualities of the stainless steels. Some stainless steels can be heat-treated and others cannot. These steels can be divided into three basic groups: austenitic, martensitic, and ferritic.

NONFERROUS METALS

Nonferrous metals are metals which do not contain iron other than small amounts as impurities. These metals include copper, zinc, tin, nickel, brass, bronze, aluminum, lead, titanium, magnesium, and many others.

Aluminum

Aluminum is one of the most widely used nonferrous metals today. It ranges from soft in its pure state to stronger than structural steel, pound for pound when alloyed. Aluminum is extremely strong and corrosion resistant under most conditions. The alloys are lighter weight than most other commercially available metals. Aluminum is offered by suppliers in all the standard shapes and sheets. Many special extruded shapes and castings are also readily available.

Magnesium

Magnesium alloys are the lightest weight of the structural metals, yet they have high strength. They are used in missles and aircraft or any application requiring high strength and light weight.

> **Caution:** Magnesium chips may ignite from heat generated during machining causing injury to the worker. Do not use water or commercial fire extinguishers to put out the fire. Smother the fire with asbestos, sand, or cast-iron turnings.

Titanium

Titanium material is as strong as steel but only half as heavy. It is extremely corrosion resistant.

Copper-Base Alloys

The following nonferrous metals are of copper-base alloys:

- Brass is an alloy of copper and zinc. It is used mostly in applications where rust and corrosion problems exist.
- Bronze is harder than brass and more expensive. It is an alloy of copper and tin. It is used in many sleeve-bearing applications.
- Copper is a base metal (a pure, metallic element). It is one of the oldest metals known. The main use of copper is for soldering iron points and electrical parts.
- Beryllium copper is a newer copper alloy, extremely tough, and long wearing. This makes it useful for high-pressure bearing applications.

> **Caution:** Machining beryllium copper is a health hazard. Do not breathe the dust or get cut on the chips of the metal.

Copper-base materials are available in rod, bar, tube, wire, strip, and sheet forms. The structural shapes of angle, channel, and beams are available. They are used for wear, corrosion resistance, and appearance applications.

HIGH-TEMPERATURE METALS

High-temperature metals and rare metals are being used more often in space exploration and national-defense projects. The high-temperature metals include those super alloys of nickel-base, molybdenum, tantalum, and tungsten.

RARE AND PRECIOUS METALS

Rare and precious metals such as gold, platinum and silver are being used in aerospace applications. Aluminum, less than a century ago, was considered a rare metal. Today, it is one of our most commonly used metals.

NONMETALLICS

Nonmetallics are also available in castings and many of the same shapes as steel bar. Many special plastic *extruded shapes* (material being forced through an opening causing its shape to conform to the opening) are now available for use. Plastics, carbon, rubber, ceramic, and wood are other non metallic materials available to the designer for use in many mechanisms. Due to the friction-free, lightweight, and corrosion-resistant features of nonmetallics, they are rapidly becoming used in machinery and aerospace applications. These materials are identified by color and weight and have the trade name of the material type printed on the surface for easy identification. When in doubt, consult your supplier.

REVIEW QUESTIONS

A.   Multiple Choice

1. Which iron is normally used for making large pipes and large, heavy machine parts?
   a. White cast iron
   b. Malleable iron
   c. Ductile cast iron
   d. Gray cast iron

2. Which of the following steels would be best to use to make cutting tools?
   a. Low-carbon steel
   b. Medium-carbon steel
   c. Cold-rolled mild steel
   d. High-carbon steel

3. The strength of a steel and its ability to hold a cutting edge is largely determined by the content of:
   a. silicon.
   b. sulfur.
   c. carbon.
   d. phosphorus.

4. Steel is an alloy of:
   a. silicon and carbon.
   b. manganese and carbon.
   c. sulfur and carbon.
   d. iron and carbon.

5. Which of the following is high-carbon steel produced in shorter lengths and used for making smaller tools?
   a. Carbon rod
   b. Drill rod
   c. Steel rod
   d. Iron rod

6. High-speed steel is a common:
   a. alloy steel.
   b. carbon steel.
   c. drill rod.
   d. wrought iron.

B. Short Answer

7. Brass is an alloy of ____________ and ____________ .
8. List three types of nonferrous materials.
9. List the four general types into which metals are divided.
10. List the three basic forms of metals used in the shop.
11. List the two common ferrous metals used in the shop.

ACTIVITY

1. From the sample material shapes provided by your instructor:
   a. Correctly identify each shape.
   b. Demonstrate the correct method to measure each shape.
   c. Give the correct size measurement of each shape.

UNIT 3 MATERIAL IDENTIFICATION AND STORAGE

OBJECTIVES

After completing this unit, the student will be able to

- explain the two numerical, steel identification systems used in the United States.

- identify steels by various shop tests.

- state the steps to correctly select materials specified on the print.

INTRODUCTION

Years ago, steel was only available in wrought iron and carbon steels. Therefore, the selection and identification was not nearly as complicated as it is today. The manufacturing industry gradually needed stronger steels to meet ever-growing design needs. New alloy steels were developed and are now in use. Correctly separating, identifying, and selecting these various metals, is a very important task. For further information refer to the Material Identification Chart in the Appendix.

METHODS OF IDENTIFYING STEELS

There are three methods of identifying steels. These methods include the two numbering systems, color-code marking systems, and various shop tests.

Numbering Systems

The two existing numbering systems were developed by the American Iron and Steel Institute (AISI) and the Society of Automotive Engineers (SAE).

The designation of steels by the SAE system is as follows:

Type of Steel	Number Series
Carbon (basic)	1XXX
Free cutting (carbon)	11XX
Free cutting (carbon)	12XX
Manganese (carbon)	13XX
Nickel	2XXX
Nickel chrome	3XXX
Molybdenum	4XXX
Chromium	5XXX
Chromium - Vanadium	6XXX
Tungsten	7XXX
Nickel - Chromium molybdenum	8XXX
Silicon - Manganese	9XXX

Example: The following example explains the SAE numbering system. The steel selected has the number 2340.

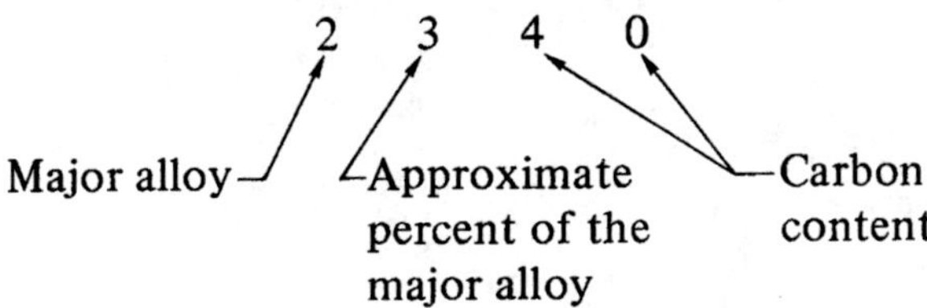

The (2) indicates the major alloy in the steel, a nickel steel. The (3) indicates the approximate percent of the major alloy used in the steel. The last two digits (40) indicate the percent of carbon in the steel. This carbon content is expressed in hundredths of a percent, as 0.40 percent, and referred to as *points of carbon*.

The AISI numerical system uses the following capital letter prefixes to identify the process used to make the steel:

A — Basic, open-hearth alloy steel
B — Acid bessemer — carbon steels
C — Basic, open-hearth or basic
     electric-furnace carbon steels
D — Acid, open-hearth steels
E — Electric-furnace alloy steels

Example: E 2340 indicates that the steel was made in an electric furnace.

Color-Code Marking System

Another steel identification system used by steel manufacturers is to color code the ends of different materials with paint. The main disadvantage with this system is that each manufacturer has its own system of coding.

When cutting bar stock, it is a good practice to cut material off the unpainted end of the bar. This allows the unused material to be identified the next time it is needed.

If a particular supplier's color-code system is used, the supplier must be known and its color code used. Always remember to cut from the shortest bar available and return any unused material to the proper storage area. Unmarked steel is not easily identified by outward appearance.

Shop Tests

The following are descriptions of various shop tests that are used to identify steel.

Spark Test. This test is done by lightly pressing the metal against the face of a revolving grinding wheel and noting the color and shape of the sparks. Refer to Figure 3-1. The spark patterns result from the small particles of metal being heated to red and yellow and then thrown into the air where they *oxidize* (burn as they come in contact with oxygen). Carbon in the metal burns rapidly. This results in a bursting of the metal particles.

The iron-carbon basic spark test is identical with the straight-iron carbon steels. As the carbon content increases, the spark explosions increase. The length of the spark flight decreases and the color of the sparks gets brighter and brighter. These spark bursts vary in color, size, shape, intensity, and the distance they travel from the grinding wheel. These characteristics depend on the alloys in the metal being ground.

For the spark test to be effective, the person testing should be able to correctly identify the type of steel being tested. A good practice is to have a sample display of known metals on hand, Figure 3-2. Compare these known sparks with the sparks of the steel being identified. Four characteristics of the spark stream that indicate the content and condition of the metal are:

- Color of spark stream
- Length of spark stream
- Number of explosions (spurts or sprigs)
- Shape of explosions (forked or repeating).

Before doing a spark test, make sure the grinding wheel has been properly dressed. Grinding wheels become clogged with metal after using them several times. The best tool to use to dress a wheel is the *diamond-pointed wheel dresser*. The rod is held against the face of the wheel. The grinding wheel is properly dressed by moving the rod back and forth across the face of the wheel.

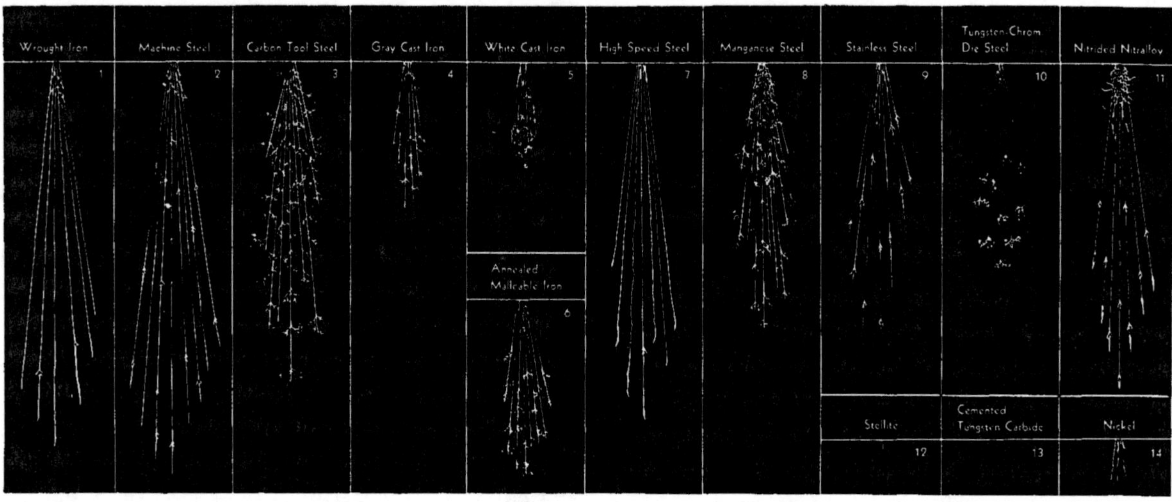

Metal	Volume of Stream	Relative Length of Stream Inches †	Color of Stream Close to Wheel	Color of Streaks Near End of Stream	Quantity of Spurts	Nature of Spurts
1. Wrought iron	Large	65	Straw	White	Very few	Forked
2. Machine steel (AISI 1020)	Large	70	White	White	Few	Forked
3. Carbon tool steel	Moderately large	55	White	White	Very many	Fine, repeating
4. Gray cast iron	Small	25	Red	Straw	Many	Fine, repeating
5. White cast iron	Very small	20	Red	Straw	Few	Fine, repeating
6. Annealed mall. iron	Moderate	30	Red	Straw	Many	Fine, repeating
7. High speed steel (18-4-1)	Small	60	Red	Straw	Extremely few	Forked
8. Austenitic manganese steel	Moderately large	45	White	White	Many	Fine, repeating
9. Stainless steel (Type 410)	Moderate	50	Straw	White	Moderate	Forked
10. Tungsten-chromium die steel	Small	35	Red	Straw*	Many	Fine, repeating*
11. Nitrided Nitralloy	Large (curved)	55	White	White	Moderate	Forked
12. Stellite	Very small	10	Orange	Orange	None	
13. Cemented tungsten carbide	Extremely small	2	Light Orange	Light Orange	None	
14. Nickel	Very small**	10	Orange	Orange	None	
15. Copper, brass, aluminum	None				None	

†Figures obtained with 12″ wheel on bench stand and are relative only. Actual length in each instance will vary with grinding wheel, pressure, etc.
*Blue-white spurts. **Some wavy streaks

Fig. 3-1 Characteristics of sparks generated by the grinding of metals (Norton Company)

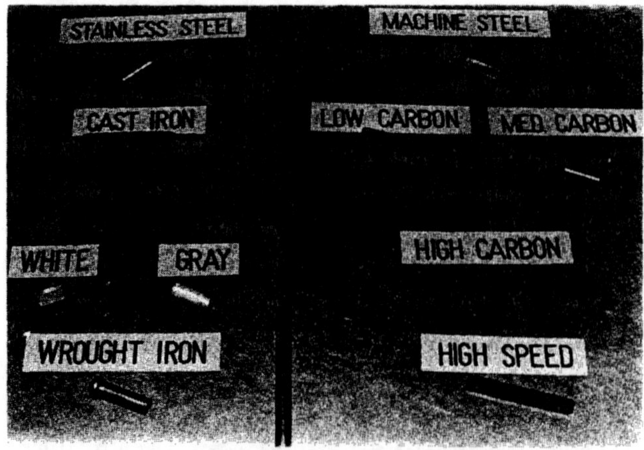

Fig. 3-2 Sample display of known metals

> **Caution**: Make sure the spark arrestor and the tool rest are correctly adjusted when attempting the spark test. Always use goggles and a shield.

Magnet Test. All ferrous metals, such as iron and steel, are *magnetic* (attracted by magnetism). Nickel, which is a nonferrous metal, is nonmagnetic. Many stainless steels and Monel Metals® are nonmagnetic. Some types of stainless steels are magnetic.

Hardness Test. This test is done with a hardness tester machine. It determines the hardness of the material. Each kind of material has a different degree of hardness.

Ring Test. Tapping metal with a small steel hammer is an easy method to identify heat-treated steels from annealed steels. Heat-treated steel has a definite ring, while annealed steel has a duller sound. High-alloy aluminum rings, while pure base aluminum has a dull sound.

Scratch Test. With this test, one material is scratched against another. One material scratches the other; thus, it is harder. An example of this is steel against aluminum.

File Test. This test can be used to determine the approximate hardness of a piece of metal using a small, smooth file.

Example: Mild steel — the file bites and cuts easily.

> Medium steel — the file bites with some pressure.
>
> High-carbon steel — it is difficult to make the file bite.
>
> Hardened Tool steel — the file does not bite, it only slides over the metal. The file teeth become dull.

Materials which the file does not mark are as hard as the file. These materials are known as *file hard*.

Chipping Test. This test is done with the use of a hammer and chisel. Various chip forms are produced from different types of materials. Low-carbon steel has a more continuous chip. Harder steel or cast iron chips have a more brittle, broken type of chip, Figure 3-3.

In some shops, various electronic computer, acid, and chemical tests are made for very close metal identification.

PROPERLY SELECTING AND CUTTING MATERIALS

Many years of experience are required to make material selection changes. A change in the kind of material may require a change in the part design. The material must meet all the specifications required by the part. This insures personal safety and mechanical success.

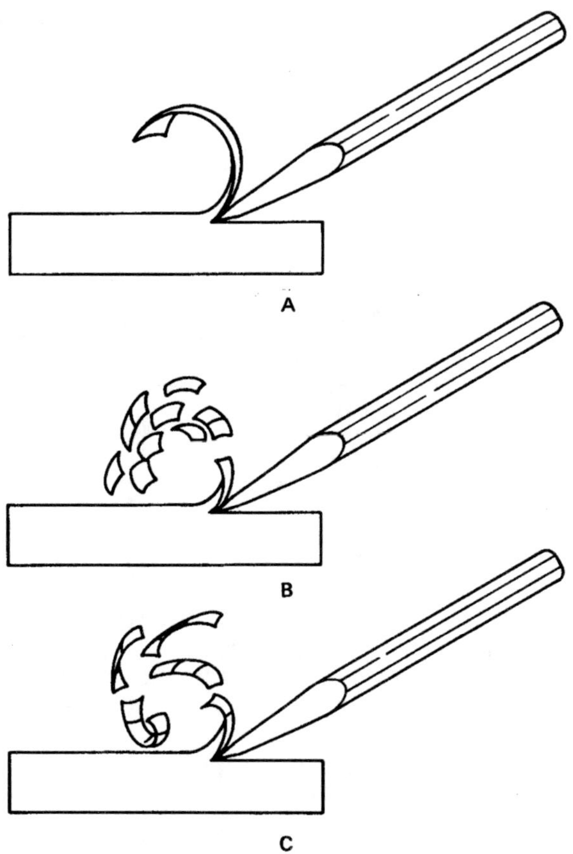

Fig. 3-3 Chip shapes from the chipping test: (A) Low-carbon steel, (B) cast iron, and (C) high-carbon steel.

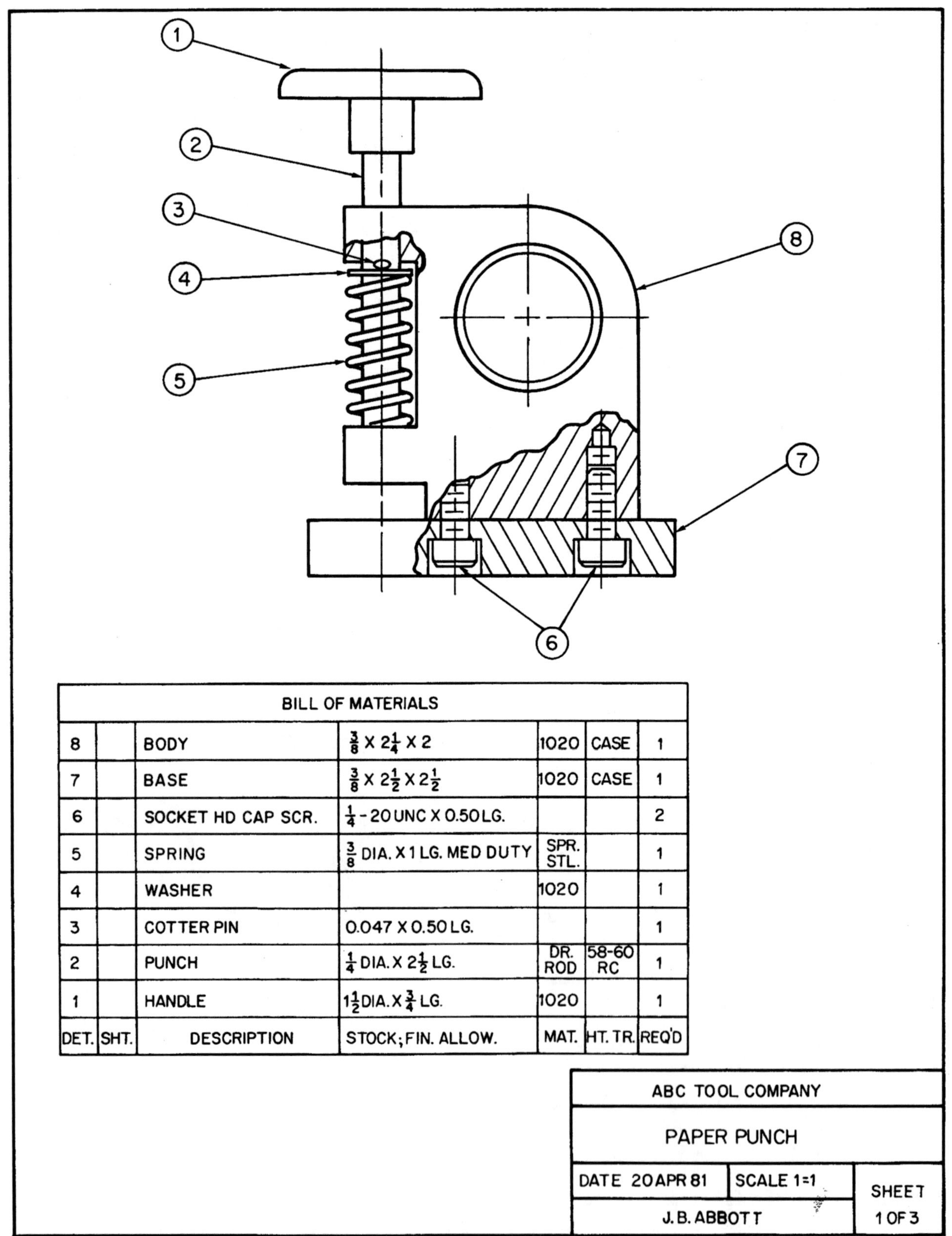

		BILL OF MATERIALS				
8		BODY	$\frac{3}{8} \times 2\frac{1}{4} \times 2$	1020	CASE	1
7		BASE	$\frac{3}{8} \times 2\frac{1}{2} \times 2\frac{1}{2}$	1020	CASE	1
6		SOCKET HD CAP SCR.	$\frac{1}{4} - 20$ UNC X 0.50 LG.			2
5		SPRING	$\frac{3}{8}$ DIA. X 1 LG. MED DUTY	SPR. STL.		1
4		WASHER		1020		1
3		COTTER PIN	0.047 X 0.50 LG.			1
2		PUNCH	$\frac{1}{4}$ DIA. X $2\frac{1}{2}$ LG.	DR. ROD	58-60 RC	1
1		HANDLE	$1\frac{1}{2}$ DIA. X $\frac{3}{4}$ LG.	1020		1
DET.	SHT.	DESCRIPTION	STOCK; FIN. ALLOW.	MAT.	HT. TR.	REQ'D

ABC TOOL COMPANY		
PAPER PUNCH		
DATE 20 APR 81	SCALE 1=1	SHEET
J. B. ABBOTT		1 OF 3

Fig. 3-4 Bill of materials

Follow the steps listed below to properly select and cut materials for a job.

1. Read the print carefully. Refer to the bill of materials shown in Figure 3-4.
2. Select the correct material.
3. Select the best shape material to complete the job quickly with minimum waste.
4. Select material of the right size, large enough to finish to the print specifications. Cut the material long enough to finish on the ends. Usually, 1/8 inch (3.1mm) to 1/4 inch (6.3 mm) excess material is left. The trueness of the material diameter depends on the accuracy of saw cuts.
5. Remember to add material for saw cuts if the piece is to be cut into shorter sections later.
6. Always select material from the shortest bars first. Most shops have an area with shorter bar remnants. Always check this area first for available stock before cutting from a new, full-length bar.

> **Caution**: Work safely to prevent possible injury. Always get help to handle long, heavy bars or use appropriate material handling devices such as cranes, forklifts, and carts.

7. Lift the load properly, Figure 3-5.
8. Always cut material from the unpainted or unmarked end.
9. Always return unused material to the proper storage racks.
10. Clean up the work area. Make good housekeeping one of your good work habits. This makes the shop a safer place to work.

MATERIAL STORAGE

Care must be taken when replacing or storing materials. When stacking materials, always use common sense. Small parts and scraps must be placed in suitable containers. Waste materials should be removed immediately.

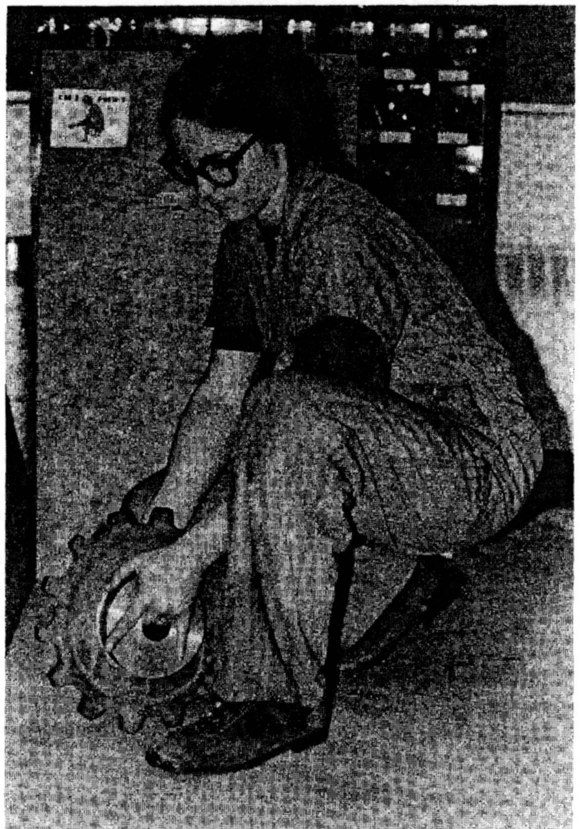

Fig. 3-5 Lift the load properly

> **Caution**: A pile or stack of materials, that slips can cause damage or lost fingers. Stack the material so it cannot fall on a worker.

In the shop, different kinds of materials are stored in separate racks. Low-carbon, hot-rolled steel in one rack and low-carbon, cold-rolled steel in another. Different shapes are also stored separately within the same rack. High-carbon steels, special steels, stainless steel, and other corrosion-resistant type materials are better stored in individual areas. Brass, bronze, plastics, and other nonmetallic material should also have special storage areas. Storing materials separately helps to maintain correct material identification. *Remember* to place the unused portions back in the correct rack.

REVIEW QUESTIONS

A.   Multiple Choice

1. Which of the following numbers is the SAE system for identifying a nickel-type steel?
   a. 1220
   b. 4340
   c. 2340
   d. 3120

B.   Short Answer

2. List four shop tests used for metal identification.

3. List the two numerical identification systems for steel used in the United States.

4. List the main disadvantage to the color-code system of steel identification.

5. List the main reason why it is so important to identify metal correctly before starting your job.

6. For a spark test to be effective, what should the worker be able to do correctly?

7. In the AISI system, if a D comes before the number 2340, this indicates that the steel was made by which process?

ACTIVITIES

1. From the sample materials provided by your instructor, correctly identify each of the following shop tests.
   a. Color code
   b. Magnetic
   c. Hardness
   d. Scratch
   e. File
   f. Chip

2. From a shop print provided by your instructor, demonstrate the basic steps to properly select and cut materials for the job.

UNIT 4 MECHANICAL FASTENERS

OBJECTIVES

After completing this unit, the student will be able to

- identify the most common mechanical fasteners used in the shop.

- state the uses of the common fasteners used in the shop.

- describe the number designation system for bolts and machine screws.

INTRODUCTION

Industrial or *mechanical fasteners* are devices that are used to locate, hold, join, or couple two or more pieces of material or work together. A standard fastener is one which is nationally interchangeable. For example, a customer could ask for a 1/2-13-UNC-2A hex cap screw SAE, grade 5. The identical product would be delivered by any standard fastener manufacturer.

There are numerous special sizes, kinds, and shapes of mechanical fasteners in use. The exploration of space has demanded increased needs for fasteners of nylon, ceramic, plastic, glass-coated metals, and new alloys, as well as new fastener shapes and designs.

Many of the parts produced in the machine shop must be assembled into machines and tools. This assembly is done at the bench with the use of many types of mechanical fasteners. Only the more commonly used fasteners are presented and discussed in this unit.

CLASSES OF FASTENERS

The four general classes of fasteners are threaded, fixed, aligning, and retaining fasteners. Adhesion and fusion bonding methods are also mentioned in this unit.

FASTENER IDENTIFICATION

Threaded fasteners are identified by the following:

- Diameter of the fastener

- Number of threads per inch

- Form of thread

- Coarse or fine thread

- Class of thread fit; #1 loose, #2 standard, or #3 tight

- External (A bolt) or internal (B nut) thread

Example 1 (English):

3/8" - 24 UNF - 2B

This threaded fastener designation is identified as follows:

3/8″ = Major diameter

24   = Number of threads per inch

U   = Unified form

NF   = National fine thread

2   = Standard class of thread fit

B   = Internal thread (nut)

Example 2 (English):

3/8″ - 16 UNC - 2B

In this example, the threaded fastener has the same major diameter (3/8″) and thread form (UN). The difference is this thread has only 16 threads per inch and according to standards is a coarse thread.

Example 3 (ISO metric):

M8 × 1.25 6A

The ISO metric threaded fastener designation is as follows:

M   = Metric symbol

8   = Size in millimetres
       (major diameter)

1.25 = Pitch in millimetres (mm)

6A   = Tolerance class symbol
       (thread fit)

UNIFIED THREADS

These 60-degree thread forms are divided into the following four series according to the number of threads per inch on a standard threaded fastener.

UNC   – Unified National Coarse

UNF   – Unified National Fine

UNEF   – Unified National Extra Fine

UNS   – Unified National Special

Coarse Threads (UNC) are used for the following reasons:

- Faster to assemble
- Easier starting with less cross threading

- Minimum assembly problems from damaged threads caused from handling
- Less chance to strip out when threaded into softer materials
- More easily tapped in brittle materials

Fine Threads (UNF) are generally used for the following reasons:

- Increased bolt strength because of greater minor diameter
- Finer adjustments
- Can be tightened tighter and remain tighter longer under vibration applications.
- Used on parts of thin wall thickness
- Easier to tap in very hard materials

Unified National Extra Fine Threads (UNEF) are used on very thin tubing and for very fine adjustment requirements.

Special Thread Series (UNS) are those of a certain number of threads regardless of diameter, such as:

- 8-thread series — high-pressure nuts and bolts and coarse thread series larger than one inch.
- 12-thread series — boiler practice and fine thread series larger than one and one-half inch.
- 16-thread series — for adjusting collars and retaining nuts.

FASTENERS SIZES AND SELECTIONS

Fasteners are available in sizes from very small up to 6-inch standards. Metric sizes of fasteners are also available. Some of the many materials used in the manufacture of fasteners are steel, alloy steel, stainless steel, brass, phospher bronze, copper, aluminum, Monel®, titanium, nylon, and plastics.

Good judgment must be used to select the best fastener for the job. The correct fastener selection insures safety in the originally designed part. Accurate assembly costs and success of the entire parts of an assembly,

Fig. 4-1 Several different fasteners are used in this special vise

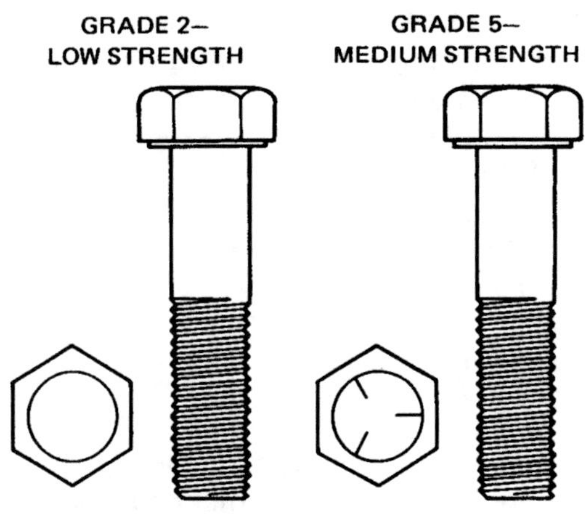

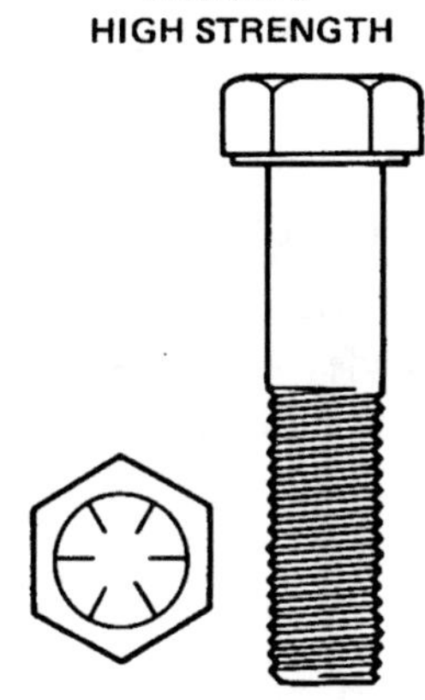

Fig. 4-2 Grading the strength of bolts (Notice the number of marks on the bolt)

machine, or tool are also benefits from proper fastener selection. Several different fastener types may be used in one assembly, Figure 4-1.

Since metric thread series and Unified National Thread series are not compatible, no metric equivalents are given.

Two standard size designations of threaded fasteners are used. If the fasteners are 1/4 inch in diameter and above, they are expressed in fractions of an inch up to about 4 inches. Below 1/4 inch in diameter, the threaded fastener is given a number from #0 to #12. A #0 screw diameter equals 0.060 inch.

To determine the diameter of a known screw size, remember that each screw size increases by an amount of 0.013 inch. Thus, a #8 machine screw has a diameter of 0.164 inch.

Example: #0 = 0.060 inch

$$\#8 = 0.060 \text{ inch} + (8 \times 0.013 \text{ inch})$$

$$\#8 = 0.060 \text{ inch} + 0.104 \text{ inch}$$

$$\#8 = 0.164\text{-inch diameter}$$

Threads are either right hand (RH) or left hand (LH). Right-hand threads are used for general-purpose applications. A right-hand thread advances when turned in a clockwise rotation. Left-hand threads are used to provide tightening to a nut due to rotation of a spindle. They are also used on one end of a turnbuckle to provide tightening of two parts.

GRADING FOR STRENGTH

The two standards used to grade the strength of bolts are the American Society for Testing Metals (ASTM) and the Society of Automobile Engineers (SAE).

The greater the number of marks on the head, the higher the quality and strength, Figure 4-2. Unmarked bolts are considered to be mild steel. Bolts of the same diameter but of different materials vary in strength. Therefore, they require different degrees of tightening.

Reference to available charts and use of special torque wrenches are required, Figure 4-3. High-strength bolts should only be used where required because of their higher costs.

> **Caution:** Use care to tighten a fastener to the proper tension so the fastener does not come loose or break off and cause personal injury.

FASTENING RULE

A threaded fastener must screw into a part at least 1 1/2 times the thread diameter, Figure 4-4. This provides maximum strength to resist stripping. *Stripping* is a pulling or pushing of the threaded fastener out of a threaded hole, thereby tearing away the thread grooves.

KINDS OF THREADED FASTENERS

Threaded fasteners are by far the most widely used style. Bolts, screws, nuts, and studs, are among the external threaded fasteners.

Bolts

A *bolt* is an external threaded fastener designed to be inserted through holes in assembled parts, Figure 4-5A. The bolt is tightened and held in place with a nut.

Machine bolts are made in two head styles, hexagon and square. Hexagon head bolts range in size from 1/4 inch to 4 inches in diameter, and in length from 1/2 inch to 30 inches. Square head bolts are made in standard sizes to 1 1/2 inches in diameter. All machine bolts are used where the parts being assembled do not require a precision bolt.

Screws

A *screw* is also an external threaded fastener, Figure 4-5B. The screw is inserted into a threaded hole and tightened or released by turning the head. These definitions do not always hold true depending on the application of the fastener.

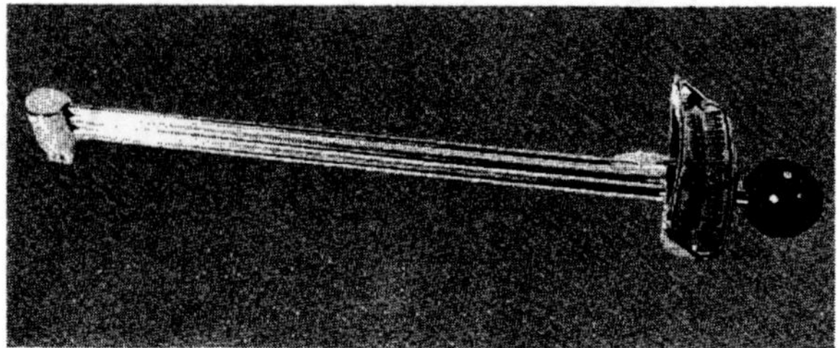

Fig. 4-3 Torque wrench

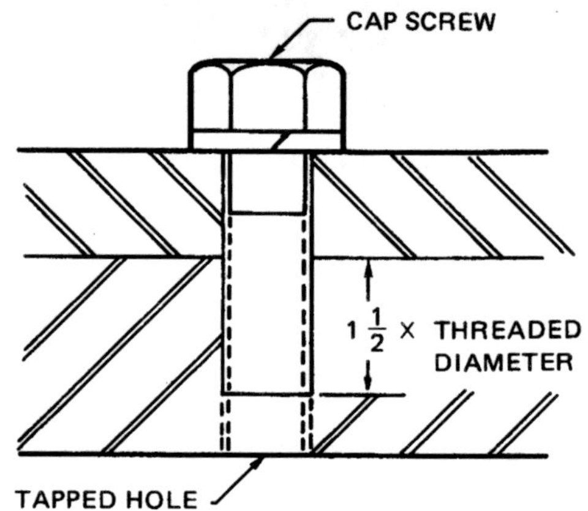

Fig. 4-4

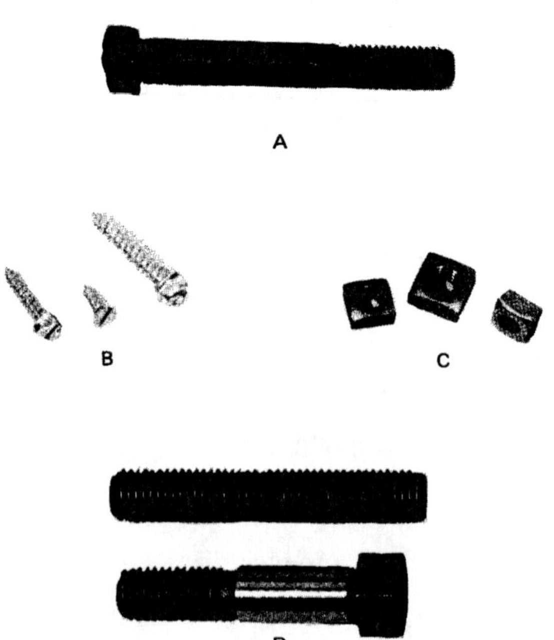

Fig. 4-5 Various standard fasteners: (A) Hex-head bolt, (B) thread-forming screws, (C) square nuts, and (D) studs

Fig. 4-6 Hex-shaped wrench being used to insert a socket set screw

Fig. 4-7 The bolt, washers, and nut are ready for assembly into a hole in the two parts behind it. As the nut is tightened, one part is clamped to the other. (Assembly is shown in front of the parts and the hole for clarity)

Cap screws are used where a stronger, more precision bolt or screw is required. They are made in the same sizes as machine bolts. *Machine screws* are made with either coarse or fine series threads. They are made in a wide range of slotted and recessed-head styles. Screwdriver or hex wrenches are used to drive these screws. Machine screws are used for general assembly work. They range in diameter from #0, which is 0.060 inch to 3/4 inch, and in lengths from 1/8 inch to 3 inches.

Set screws are used to lock one workpiece to another. The most common use is to position or hold a pulley, gear, or collar to a shaft. Many head and point styles are made in several sizes for different uses. Special wrenches are made for use with these styles, Figure 4-6. The socket shoulder screw is used as a fixed shaft about which a workpiece can rotate.

Thumb screws and wing screws are used with parts to be fastened or adjusted frequently and quickly without the use of tools. *Thread-forming screws*, self-tapping screws or thread-cutting screws, form or cut their own threads as they are screwed into the workpiece. This eliminates the need of tapping. A hole must first be made in the material to be correct size. Some of these screws are made with a short drill point on the end. This eliminates the predrilling operation. *Drive screws* are forced into the correct size hole with a hammer or suitable press. They make a permanent connection. Drive screws are often used to attach nameplates or identification plates on machines and tools.

Nuts

Nuts are internal threaded fasteners used on a bolt or screw for tightening or holding, Figure 4-5C. They are also used where it is necessary for the workpiece to be removed and reassembled. Nuts are made in as many sizes as there are bolts. Clamping is achieved as the nut advances toward the screw head, Figure 4-7.

Nuts are made in two shapes, square and hexagon.

The *square nut* is manufactured in the regular square and the heavy-duty square. Machine-screw nuts and fractional-size nuts are the two groups of square nuts. Square nuts are available in coarse thread only.

In industry, the *hex nut* is the most widely used form of nut. Nuts are identified by the bolt they fit and by their outside size. Common hex nuts are made in different outside-size styles and thicknesses. Hex nuts are made in both coarse and fine series. Hex nuts are divided into patterns of regular, heavy, or jam styles. These styles vary mainly in thickness.

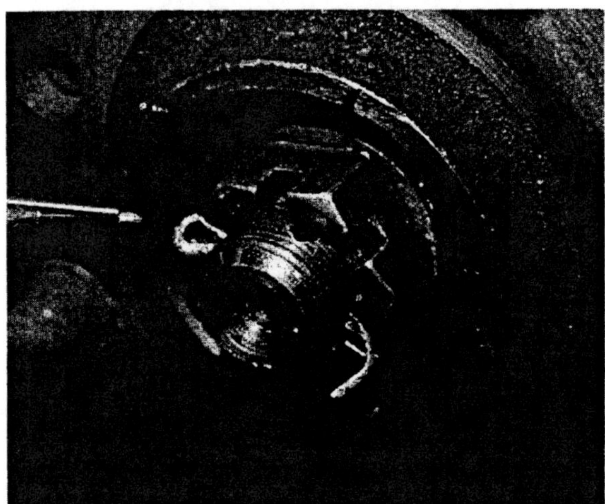

Fig. 4-8 Use of a slotted nut and cotter pin

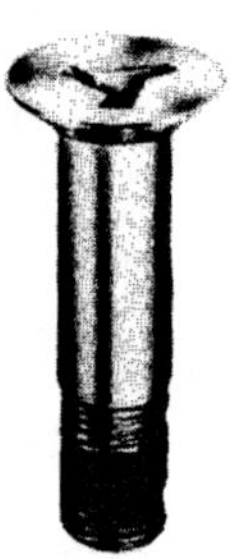

Fig. 4-9 A nylon spot in the threaded fastener resists loosening effects. (ESNA Division of Amerace Corporation)

Regular nuts are used in normal applications and are unfinished except on the thread. *Heavy nuts* are used where heavy workloads are applied. The *jam* nut is a thinner nut used to either back up or lock the main nut. A wrench is used to hold the main nut while the jam nut is firmly tightened against it.

Special Nuts

Castellated and slotted nuts have milled slots across the flats so they can be locked with a cotter pin or soft steel wire, Figure 4-8. Many automobile manufacturers make use of castellated nuts. The *elastic stop nuts and compression stop nuts* are used in severe vibration applications. A nylon collar, or *nylon spot,* in the nut or on the bolt or screw resists the loosening effects, Figure 4-9.

Cap and acorn nuts are often used where a projecting thread must be covered or protected from accidental damage or corrosion attack. *Wing nuts* are used where frequent adjustment or removal is necessary.

Studs

Studs are another commonly used threaded fastener. (Refer to Figure 4-5D.) It is a simple cylinder with threads on each end. A *stud bolt* has threads throughout its entire length. Figure 4-10 shows the application of a special threaded stud.

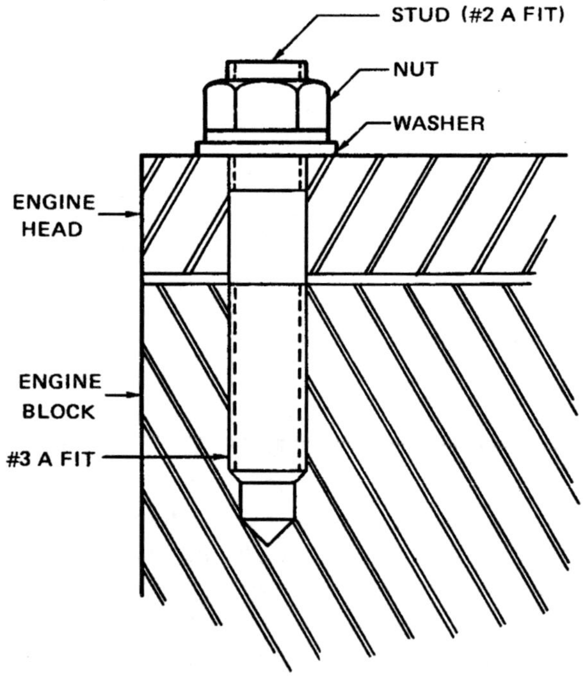

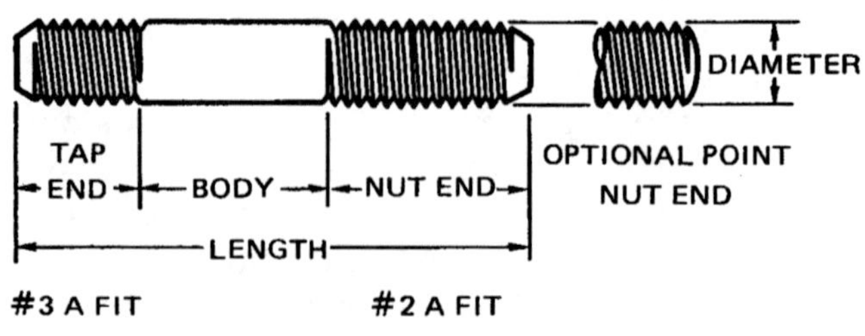

Fig. 4-10 Tap-end stud having #2A and #3A thread fits

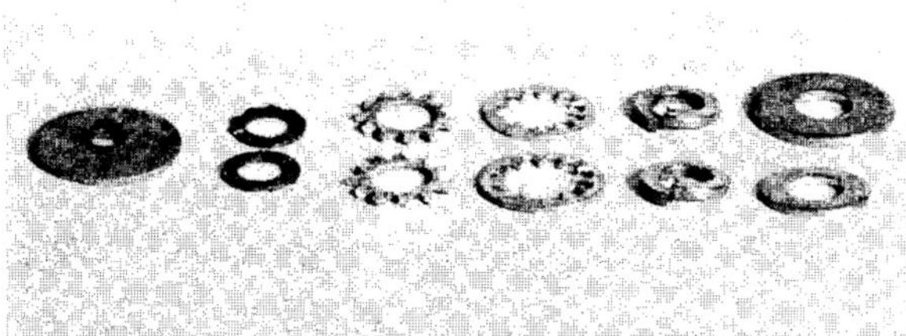

Fig. 4-11 Various styles of lock washers

**Fig. 4-12 Internal threaded insert
(Grov-Pin Corporation)**

Washers

Flat washers are discs with a clearance hole in the center to accept a certain size screw or bolt. Washers are used between a nut or bolt and the workpiece to provide a surface against which the fastener head or the bearing surface of a nut can bear. Washers are made in many different materials for various uses.

Lock washers are made in many styles, Figure 4-11. They are used to prevent a bolt or nut from loosening under rotation or vibration. Some of the materials used to make lock washers are alloy steel, stainless steel, silicon bronze, and phosphor bronze.

Special Threaded Fasteners

Internal threaded inserts, Figure 4-12, may be used where an internal thread is damaged or stripped. Various driving tools are used to place the insert into the part, Figure 4-13. The insert is screwed into a hole tapped to the same size thread as the outside of the insert   Threaded inserts are used in new aplications. They are also used in repair where stronger threaded holes are required in products made from soft plastics and metal, such as aluminum.

KINDS OF NONTHREADED FASTENERS

Fixed Fasteners

Rivets are nonthreaded fasteners used to make a more permanent assembly of parts. Rivets are made in various head shapes and with a round body. Rivets vary in size, length, and also in the type of material used to manufacture them. The rivet body should extend through the work a length of 1 1/2 times the rivet diameter, Figure 4-14. This provides a

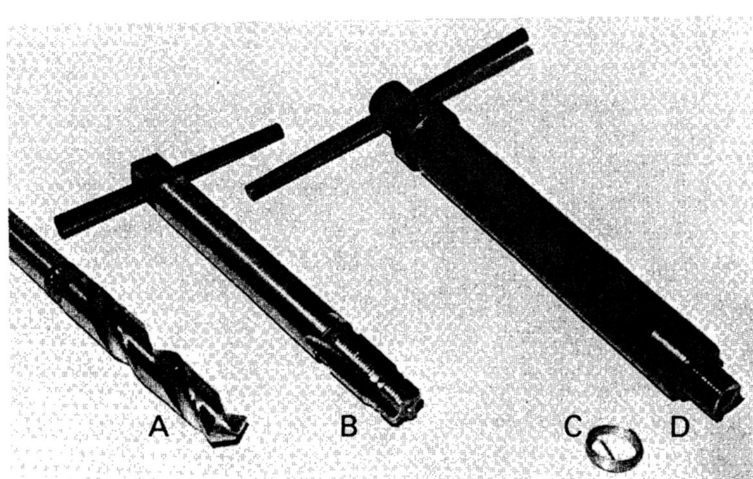

Fig. 4-13 Installation kit for threaded inserts: (A) Drill bit, (B) tap, (C) threaded insert, and (D) insert driver

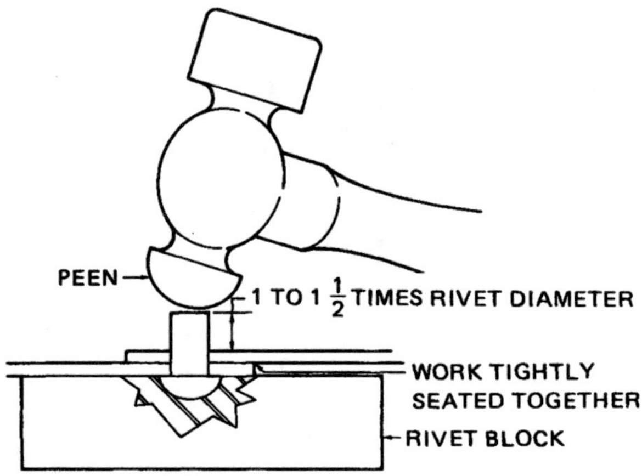

(A) STRIKE THE CENTER OF THE RIVET TO START THE HEAD FORM.

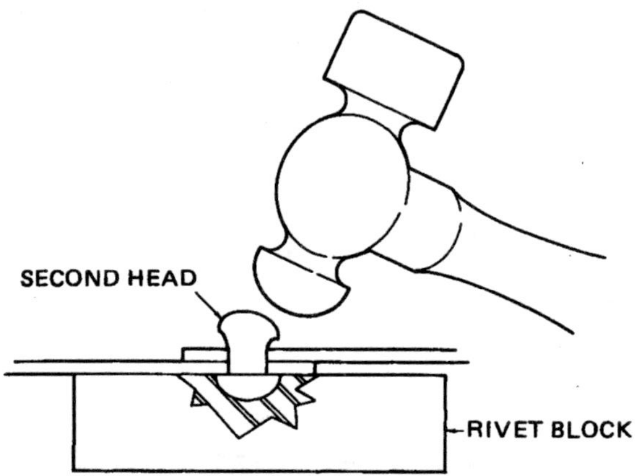

(B) ROUND THE END OF THE RIVET BY PEENING.

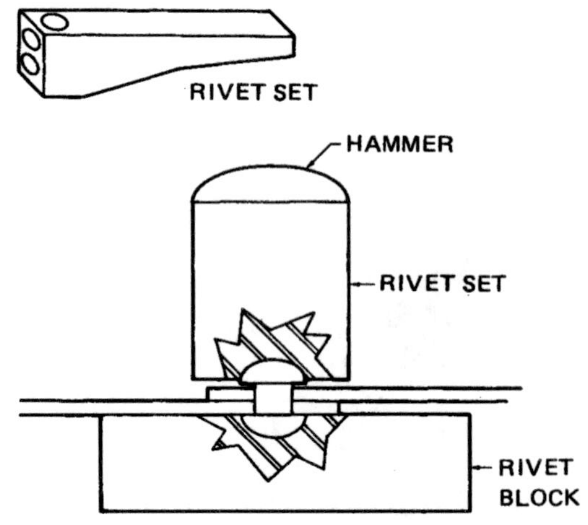

(C) FORM THE RIVET HEAD WITH THE RIVET SET.

Fig. 4-14 Steps to rivet two pieces of work together

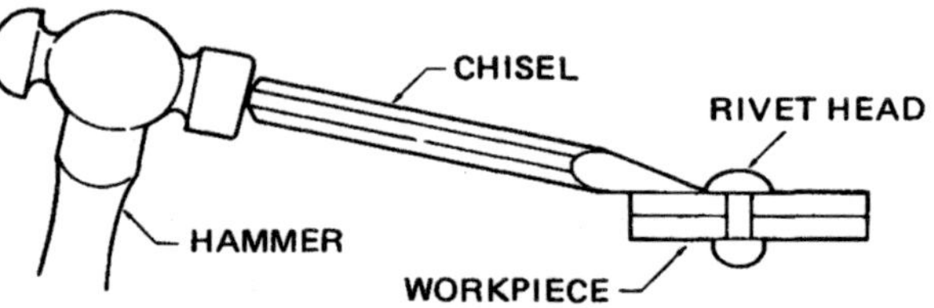

Fig. 4-15 Removing rivets with a hammer and chisel

**Fig. 4-16 Pop rivets and a riveting tool
(Marson Fastener Corporation)**

good second head. Figure 4-15 shows a rivet being removed.

A *blind rivet* provides an easier, lower cost method of riveting, than described above. A special tool is needed to install the rivet in place, Figure 4-16. This rivet is also called an *expanding* or *pop* rivet. The rivet mandrel, or pin, is inserted into the hole, Figure 4-17A. The worker squeezes the tool handles together until the pin or mandrel breaks, Figure 4-17B. These rivets are also installed with power tools for easier quicker assembly.

Aligning Fasteners

The more commonly used nonthreaded pins are described below.

Cotter pins are used to retain parts on a shaft or to lock a nut or bolt as a safety precaution, Figure 4-18A. Cotter pins make a quick assembly and disassembly possible. *Dowel pins* are made of heat-treated alloy steel. They are used in assemblies where parts must be accurately positioned and held in a fixed relation to one another, Figure 4-18B.

Taper pins are used to locate workpieces where frequent removal and assembly is

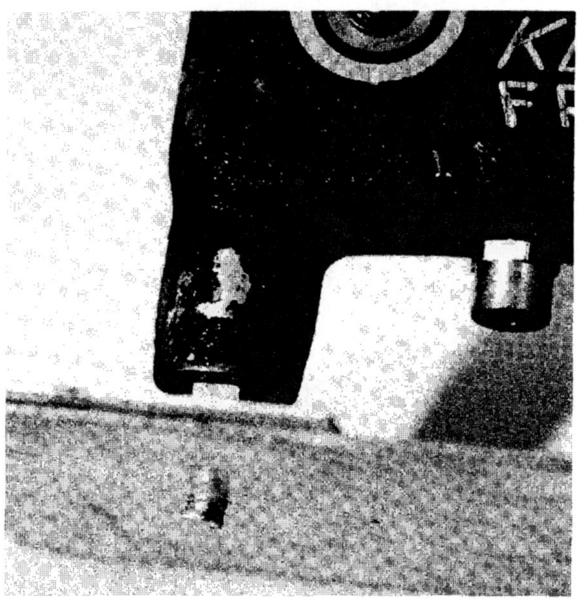

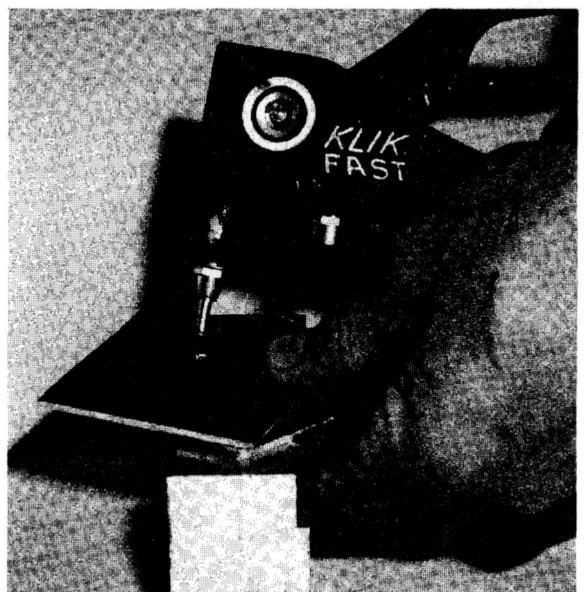

Fig. 4-17A Rivet mandrel inserted in tool ready to be inserted into hole (Marson Fastener Corporation)

Fig. 4-17B Rivet squeezed into place holding the parts together (Marson Fastener Corporation)

necessary, Figure 4-18C. *Split dowel pins* are also called *spring* or *roll pins*, Figure 18-D. This type of pin can be driven into a rough drilled hole and lock itself into the hole with its outward spring action. Split dowel pins are quick to use and can be used over and over in applications of impact, shock, or vibration.

RETAINING FASTENERS

Retaining rings, Figure 4-19, prevent the slippage or movement of one workpiece to another in a plane perpendicular to it. Retaining rings can easily be installed in machined grooves, internally in housings, or externally

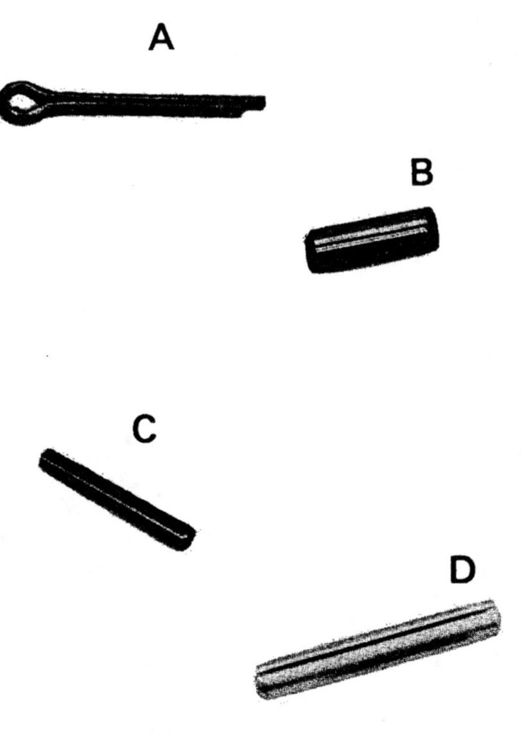

Fig. 4-18 Aligning fasteners: (A) Cotter pin, (B) dowel pin, (C) taper pin, and (D) split dowel pin

Fig. 4-19 An external retaining ring (Waldes Kohinoor, Inc.)

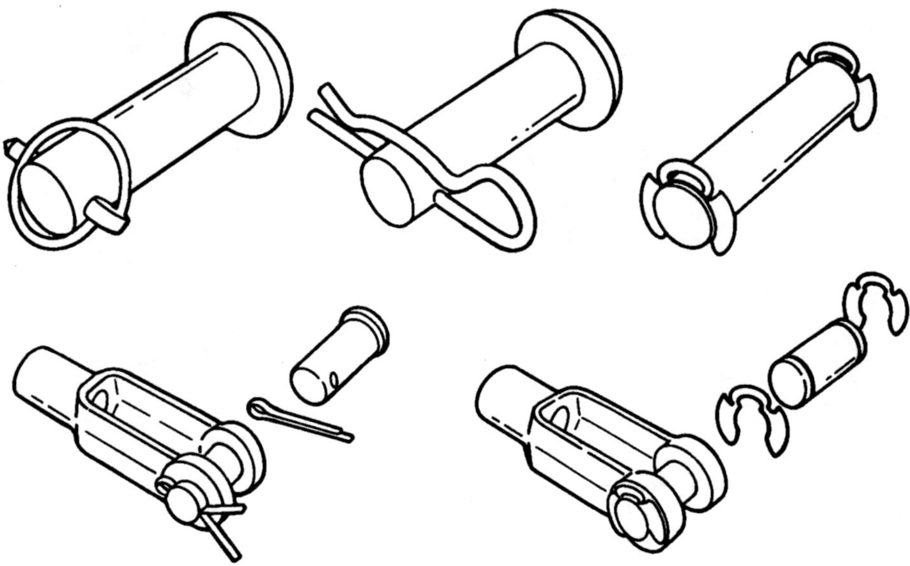

Fig. 4-20 Applications of retaining rings

on shafts or pins, Figure 4-20. Special pliers are used to rapidly install and remove retaining rings, Figure 4-21.

KEYS

Keys, Figure 4-22, are made in various shapes and materials. They are used to prevent free rotation of a gear or pulley on a shaft. The key becomes the driving member between the shaft and the pulley or gear on the shaft. Keys fit partially imbedded in the shaft and partially in the hub of the pulley or gear. The term *keyseat* refers to the slot in the shaft. A *keyway* is the slot in the hub, Figure 4-23. The following are the more commonly used keys in assembly.

Square keys have an equal width and height and are preferred on shafts up to 6 1/2 inches in diameter. Rectangular keys are recommended above this size. *Woodruff keys* are semicircular in shape. The top of the key fits into the keyway of the mating part. They are used for light loads and where their location on the shaft should remain constant. They are retained in a machined pocket.

Woodruff keys do not require a set screw to retain them in position. Woodruff keys are

Fig. 4-21 Standard pliers used on internal ring application (Waldes Kohinoor, Inc.)

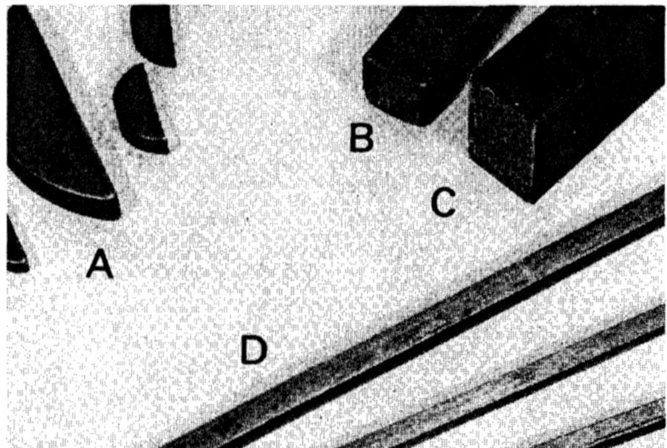

Fig. 4-22 Various key styles and materials: (A) Woodruff, (B) flat, (C) square, and (D) square key stock

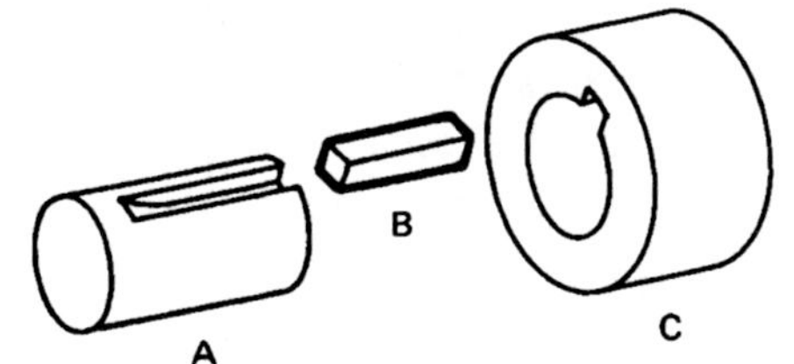

Fig. 4-23 (A) Keyseat, (B) key, and (C) keyway

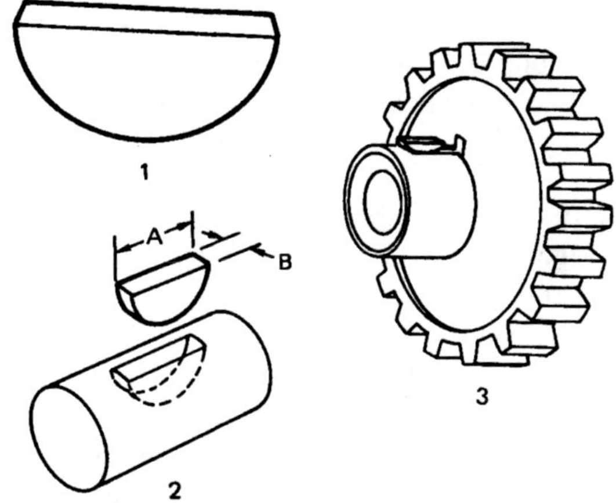

Fig. 4-24 (1) Woodruff key, (2) machined shaft, and (3) assembly

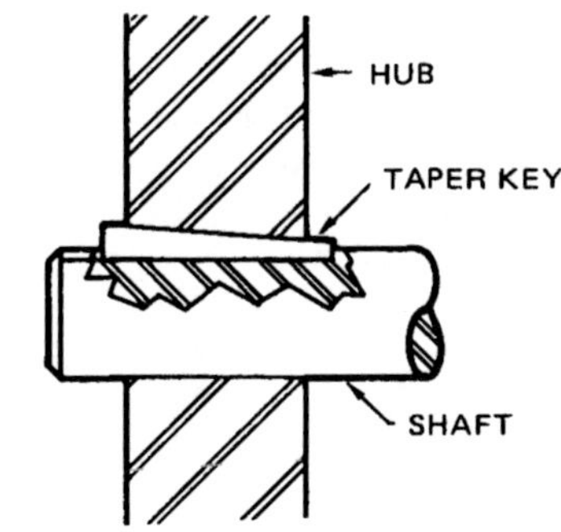

Fig. 4-25 Taper key installed in hub

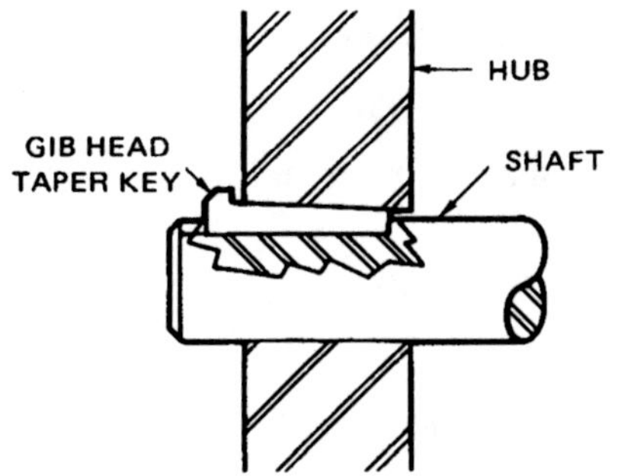

Fig. 4-26 Gib-head taper key installed in a hub

often used on taper shafts. Woodruff keys are designated by a number. The key numbers indicate nominal key dimensions. The last two digits give the nominal diameter (A) in 8ths of an inch. The digits preceding the last two give the nominal width (B) in 32nds of an inch. Refer to Figure 4-24.

Example: Woodruff Key #1012

12 − 8ths = 1 1/4-inch diameter (A)

10 − 32nds = 5/16 inch wide (B)

End-milled or pocket keyways are square keys held in place by set screws. *Taper keys*, Figure 4-25, are used in high torque or vibration loads. This key has a taper of 1/8-inch per foot. *Gib head taper keys*, Figure 4-26, are used when only one side of an assembly is accessible. A space remains between the gib head and the hub of the gear or pulley. The key is removed by driving a wedge into the space and removing the key.

A *feathered key*, Figure 4-27, is a key that is secured in a keyseat with some type of fastener. This type key is often a part of a sliding pulley or gear assembly. The *Johnson key* is another form of driving fastener device,

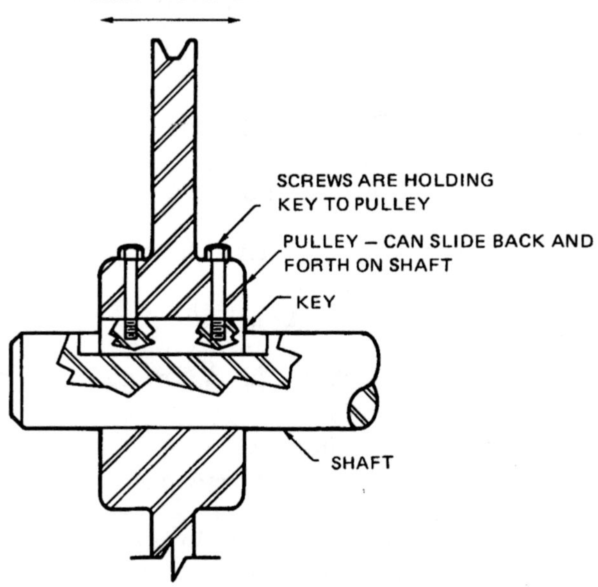

Fig. 4-27 Feathered key

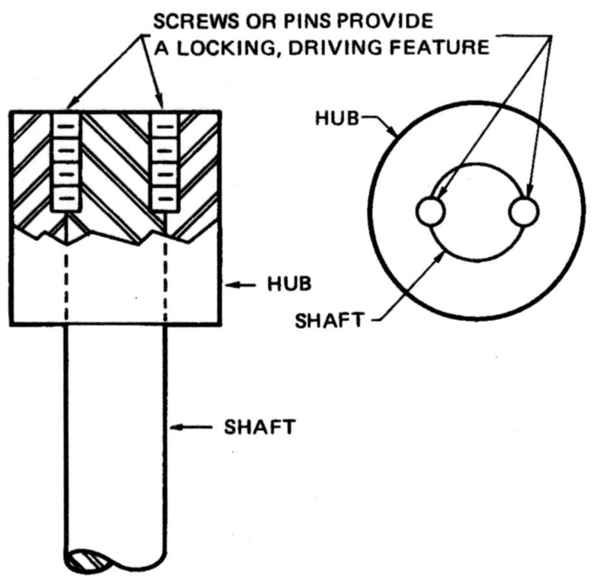

Fig. 4-28 Johnson key

Figure 4-28. Two holes for pins or screws are drilled at the intersection of the shaft and the bore of the mating parts.

OTHER FASTENING METHODS

Adhesive fastening (anaerobic machinery adhesives) is one of the newer methods of joining materials. Anaerobic adhesives and sealants refers to a liquid, paste, or gel agent activated by the removal of air. They lock parts against vibration-induced loosening and seal them against leakage and corrosion. Some advantages of using adhesives are listed below:

- The load is evenly distributed over the entire joint.
- Continuous contact between the mating surfaces provides full strength.
- Fastener holes are eliminated.
- Welding heats are not needed; thus, warping is eliminated.
- There are no external fastener heads.

Soldering is the process of fastening metals together with a nonferrous, low melting-point metal which sticks to the surfaces being joined. *Brazing* is a welding process which uses a nonferrous metal filler rod with a melting point above 1000 degrees Fahrenheit, but below the melting point of the metals being joined. *Welding* is the joining of metal parts by melting and fusing the two metals together.

FASTENER STORAGE

Fasteners must be stored properly. Clearly marked bins provide separate fastener storage. Shop time is saved by having fasteners readily available.

REVIEW QUESTIONS

A. Multiple Choice

1. Correct fastener selection is important to insure which of the following?
   a. Safety
   b. Operational success
   c. Accurate assembly costs
   d. All of the above

2. Which of the following is the most commonly used for fastening materials?
   a. Adhesives
   b. Threaded fasteners
   c. Nonthreaded fasteners
   d. Welding

3. Which of the following is the diameter of a #10 machine screw?
   a. 0.212 inch
   b. 0.190 inch
   c. 0.162 inch
   d. 0.130 inch

4. To provide maximum strength to the thread to resist stripping, a threaded fastener must screw into a part at least ___________ times the thread diameter.

   a. 1 1/8   c. 1 1/2
   b. 1/2   d. 3/8

5. A right-hand threaded fastener advances into a part by turning it in which direction?

   a. Clockwise   c. Counterclockwise
   b. Backward   d. Forward

6. Five marks on the head of a bolt would indicate that the strength of the bolt is:

   a. low.   c. high.
   b. medium.   d. very high.

7. Which of the following fasteners is made in hexagon and square head styles?

   a. Cap screw   c. Machine bolt
   b. Machine screw d. Slotted nut

8. Which of the following is most commonly used to position or hold a pulley, gear, or collar to a shaft?

   a. Stud bolt   c. Thumb screw
   b. Castellated nut d. Set screw

9. A nonthreaded fastener used to make a more permanent assembly of parts is called a(n):

   a. aligning fastener. c. lock washer.
   b. rivet.   d. dowel pin.

10. What fastener is used to prevent free rotation of a gear or pulley on a shaft?

    a. Key   c. Lock washer
    b. Cotter pin   d. Flat washer

B. Short Answer

11. Identify the following items for the threaded fastener designation given below:

$$5/8'' - 11 - UNC - 2B$$

    a. Internal thread   ________________
    b. Outside diameter   ________________
    c. Thread form   ________________
    d. Thread series   ________________
    e. Threads per inch   ________________
    f. Thread fit   ________________

12. List three common uses for fine threaded fasteners.

13. When a threaded fastener has been *stripped*, what does this mean?

14. Name two advantages of using adhesive fasteners.

15. A simple cylinder with threads on each end is called a ____________.

16. List the four general classes into which fasteners are grouped.

ACTIVITY

1. From a selection of fasteners provided by the instructor:
    a. Identify the type of fastener and give its correct name.
    b. Identify the hardness grade of each fastener.
    c. Identify the kind of material of each fastener.
    d. Give a common use for each fastener.

SECTION II
MEASUREMENT SYSTEMS, TOOLS, AND FITTING
UNIT 5 UNITS OF MEASURE

OBJECTIVES

After completing this unit, the student will be able to

- identify the common linear, fractional, decimal, and metric systems of measuring.

- explain the purposes for measurements as related to interchangeability.

INTRODUCTION

Before modern-day industrial operations, an individual craftsperson was often responsible for completely making a product. Since all the parts were made and assembled by the one worker, they only needed to conform to that individual's system of measurement.

As machines replaced people and mass production of parts began, parts needed to be made *interchangeable* (replacement parts must fit with the same accuracy as the original part). This led to the need for standards in measuring. Total standardization of measurement has not yet been set up throughout the world. Most measurement now conforms either to the English or the metric system. Machinists must begin to think and work in terms of metric measurement. In the future, our measuring system will be entirely metric.

Once systems and units of measure are carefully understood, they can then be applied to the degree of accuracy required through the use of measuring tools.

BASIS OF PRECISION MEASUREMENTS

The basic inch and basic metre standards are based on a very accurate wave length of light.

Example:

1 wave length	=	0.0000238 inch
1 inch	=	42,016.807 wave lengths of light
1 metre	=	1,650,763.73 wave lengths of light

Light waves do not change much with temperature and atmospheric conditions. Therefore, this precise measurement can be done in all parts of the world. This standard unit for linear measurement makes it possible to obtain precision measurement of mass-production methods and interchangeability of parts.

LINEAR MEASUREMENT

In the shop, the most common measurements made are of length called *linear (straight-line) measurement*. The length, width, thickness, and diameter of parts all require linear measurement. Several English and metric measuring tools are available for linear measurement.

SYSTEMS OF MEASUREMENT

The two systems of linear measurement commonly used in the United States are the English system and the metric system.

English System

The English system of measurement uses the *yard* as the basic unit of length. In machine shop work, the more commonly used English unit is the *inch*. Other common multiples of the inch are:

12 inches = 1 foot
36 inches = 1 yard

The inch may be divided into smaller parts by either common fractional or decimal divisions.

Fractional divisions of an inch are found by dividing the inch into equal parts, Figure 5-1. The more frequently used divisions are:

halves	= 1/2	sixteenths	= 1/16
quarters	= 1/4	thirty-seconds	= 1/32
eighths	= 1/8	sixty-fourths	= 1/64

When smaller units of measure are required, decimal divisions are used, Figure 5-2. The inch is divided into the following:

tenths	= 0.1 or 1/10
hundredths	= 0.01 or 1/100
thousandths	= 0.001 or 1/1000
ten-thousandths	= 0.0001 or 1/10,000

The smallest part of an inch the machinist needs to measure is one ten-thousandths inch (0.0001 inch).

In shop work, it is common practice to use fractions of an inch expressed in decimals. These are called *decimal equivalents* of an inch. The commonly used fractions of an

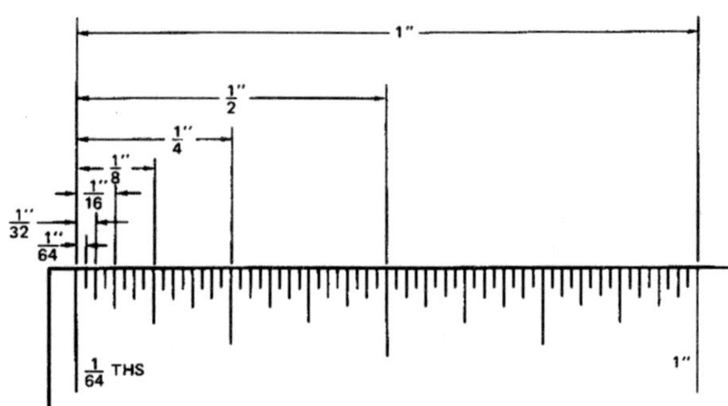

Fig. 5-1 Divisions of the inch with common fractions

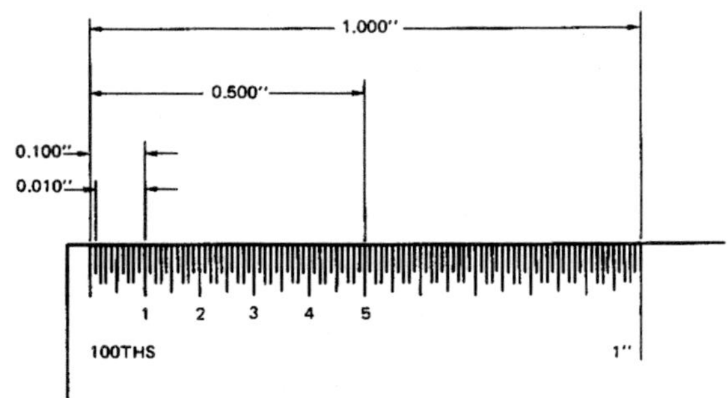

Fig. 5-2 Divisions of the inch with common decimals

inch and their decimal equivalents are listed in the Appendix. They should be learned and memorized through use.

Metric System

The metric system (Systeme International, or SI) has the *metre* as the basic unit of length. The metre (39.37 inches) roughly corresponds to the yard (36 inches). The SI system is the means by which the metric system is becoming standardized.

All units of the metric system are divided into ten parts. All subdivisions are either divided by or multiplied by ten. The following are commonly used metric length units:

0.001 metre	= 1 millimetre (mm)
0.01 metre	= 1 centimetre (cm)
0.1 metre	= 1 decimetre (dm)
1.00 metre	= unit metre (m)
10.00 metres	= 1 dekametre (dam)
100.00 metres	= 1 hectometre (hm)
1000.00 metres	= 1 kilometre (km)

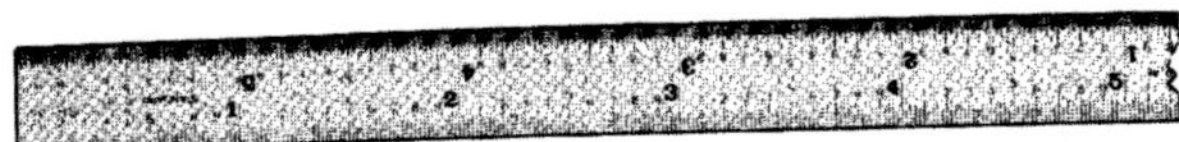

Fig. 5-3 Metric rule graduated in millimetres and half millimetres (Swiss Precision Instruments, Inc.)

1 mm = 0.001 m

1 cm = 10 mm

1 dm = 10 cm or 100 mm

1 m   = 10 dm or 100 cm or 1000 mm

1 km = 1000 m

1 mm = 0.03937 in

1 cm = 0.3937 in

1 m   = 39.37 in

Metric rules are graduated to measure in millimetres and half millimetres, Figure 5-3. In the shop, metric measuring is mainly expressed in metre (m) and millimetre (mm) length units. Fractions of the millimetre are expressed in decimals (i.e., 0.02 mm and 0.050 mm). Linear metric dimensions are always expressed in multiples and submultiples of the metre (i.e., 8.731 mm and 22.225 mm).

ENGLISH AND METRIC CONVERSIONS

Conversion between the two systems of SI and English is done mathematically or by reference to conversion charts (see Appendix). Many electronic calculators are now available to convert directly from system to system. When available, the calculator or the conversion charts should be used to save valuable shop time. The mathematical conversion is made by memorizing and applying the basic conversion factors.

Example: 1 yard (36 inches) $= \dfrac{3600}{3937}$ metre

This fraction would not be easily used in conversion calculators. The decimal form of this fraction equals 0.914 40 metre. Therefore,

1 yard (36 inches) = 0.914 40 metre

1 inch = 1/36 of 0.914 40 metre
  or 1/36 X 0.914 40 metre

Therefore, $\dfrac{0.914\ 40}{36}$ = 0.025 400

Since 1 metre = 1000 millimetres,

1 inch = 0.025 400 X 1000 millimetres

1 inch = 25.400 0 millimetres

This conversion factor or constant, 1 inch = 25.4 millimetres, is common and easily memorized. To find inches knowing millimetres, simply divide the known millimetre by 25.4.

Example: Convert 1000 millimetres to inches.

$$\dfrac{1000}{25.4} = 39.37 \text{ inches}$$

1000 millimetres = 39.37 inches

This arithmetic can further be simplified by converting with multiplication. Instead of dividing the millimetre by the constant 25.4 millimetres to find inches, multiply by the reciprocal of 25.4 which is 1/25.4 or 0.039 37. Therefore, in the previous example, 1000 millimetres = 1000 X 0.039 37, or 39.37 inches.

Example: Convert 0.898 inch to millimetres.

Knowing inches, multiply inches by 25.4 to find millimetres. Thus,

0.898 X 25.4 = 22.8 millimetres

0.898 inch   = 22.8 millimetres

The following conversion formulas may be used to obtain approximate length equivalents from English to metric and metric to English:

- Inches   = millimetres X 0.04
- Inches   = centimetres X 0.4
- Yards   = metres X 1.1
- Millimetres = inches X 25
- Centimetres = feet X 30
- Metres   = yards X 0.9

REVIEW QUESTIONS

A.   Multiple Choice

1. One metre equals which of the following?
    a. 25.4 inches
    c. 39.37 inches
    b. 3.937 inches
    d. 254 inches

2. Which of the following is the smallest part of an inch generally measured in the shop?
    a. 0.100 inch
    c. 0.001 inch
    b. 0.010 inch
    d. 0.0001 inch

3. When converting the two systems of measurement, it is best accomplished by:
    a. math calculations.
    c. memory.
    b. conversion charts.
    d. asking a fellow worker.

4. Most measurements in the shop are which of the following?
    a. Angular
    c. Round
    b. Circular
    d. Linear

5. One yard equals which of the following?
    a. 18 inches
    c. 24 inches
    b. 12 inches
    d. 36 inches

B.   Short Answer

6. List the two systems of measurement used in the shop.

7. List the four smaller units of measure commonly used in the inch decimal system.

8. Define the term *interchangeable* in relation to parts.

C.   Matching

9. Match the correct decimal equivalent in Column 2 with the fractions in Column 1.

Column 1	*Column 2*
a. 1/16	1. 0.875 inch
b. 1/8 inch	2. 0.0625 inch
c. 3/16 inch	3. 0.375 inch
d. 1/4 inch	4. 0.500 inch
e. 5/16 inch	5. 0.3125 inch
f. 3/8 inch	6. 0.750 inch
g. 1/2 inch	7. 0.250 inch
h. 5/8 inch	8. 0.125 inch
i. 3/4 inch	9. 0.625 inch
j. 7/8 inch	10. 0.1875 inch
k. 1/64 inch	11. 0.032 inch
l. 1/32 inch	12. 0.0156 inch

10. Using the rule learned to convert inch measurements to metric, match
    the following inch measurements in Column 1 with the correct metric
    measurements in Column 2.

    Column 1   *Column 2*
    a. 5/8 inch   1. 6.35 millimetres
    b. 1/8 inch   2. 25.4 millimetres
    c. 1/4 inch   3. 15.875 millimetres
    d. 1 inch   4. 3.175 millimetres

ACTIVITIES

1. From conversion tables provided by your instructor, identify various
   metric to inch and inch to metric conversions.

2. On steel rules provided by your instructor, demonstrate the ability to
   accurately read various given dimensions.

UNIT 6 SEMIPRECISION MEASURING TOOLS AND GAGES

OBJECTIVES

After completing this unit, the student will be able to

- identify the semiprecision measuring tools commonly used in the shop.
- accurately read the calibrations on shop measuring tools, including metrics.
- demonstrate the proper use and care of measuring tools.
- identify direct and indirect-reading tools.

INTRODUCTION

Accuracy refers to a measurement being within certain required dimensions. In the past, measuring tools did not demand great accuracy. Most products were custom made by hand. Satisfactory operation of an assembly of parts did not depend on very close fits.

With the demand for higher speeds and longer lasting parts came the need for closer measuring and fitting. The benchworker must thoroughly understand the operations of required measurement and the tools to obtain these measurements This unit discusses the general methods of measuring and semiprecision measuring tools.

TOOL SELECTION AND CARE

Correct selection of a measuring tool and its proper care and use are very important to quality workmanship. When using measuring tools, observe the following:

- Avoid scratches or nicks that cover graduations or distort their contact surfaces.
- Protect tools against moisture. Rust pits and destroys finely finished surfaces.
- Wipe the fingerprints off tools after using.
- Keep tools free from dirt or damage by storage in separate boxes or cases, Figure 6-1.
- Apply a light dressing of oil with a soft, lint-free cloth. This protects tools in storage, Figure 6-2.
- Measuring tools must be set or calibrated to zero to a known standard, Figure 6-3.

MEASURING METHODS

Linear measurements are perhaps the most common measurements made in general shop practice. These measurements are made either directly or indirectly.

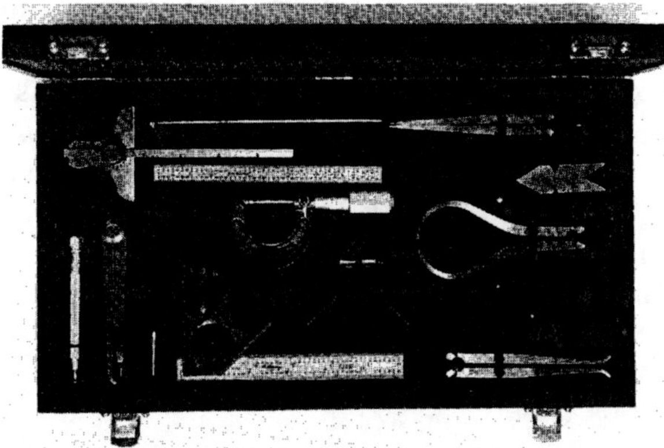

Fig. 6-1 Set of tools arranged in a convenient case (MTI Corporation)

Direct measurements are read directly from a graduated scale on the tool. The tool is in contact with the workpiece, Figure 6-4. Graduated rules are direct reading tools.

Indirect measurements are made by comparison with a standard or direct reading tool, Figure 6-5, and these tools do not read directly upon them. The measured distance must be transferred to a direct-reading measuring tool to determine a readable measurement size. These tools include calipers, telescope gages, and small hole gages.

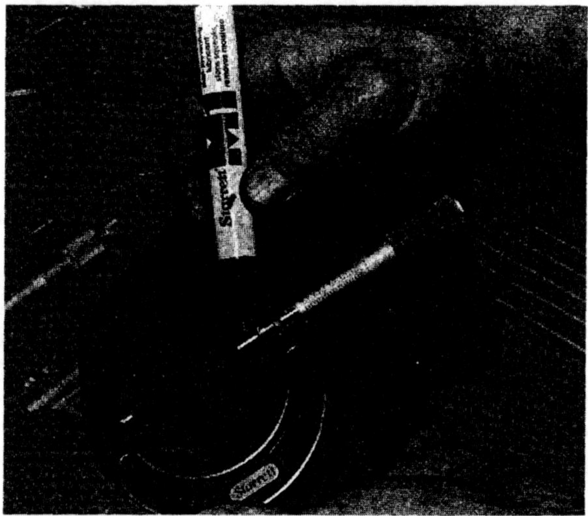

Fig. 6-2 Properly clean and lightly oil tools to be stored. (L. S. Starrett Company)

Fig. 6-4 Direct reading of a part taken by a graduated rule (L. S. Starrett Company)

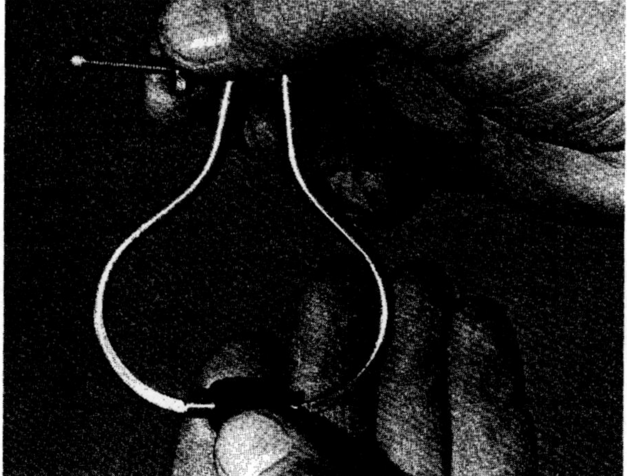

Fig. 6-3 Indirect-reading outside caliper being set to a known standard for comparison measurement

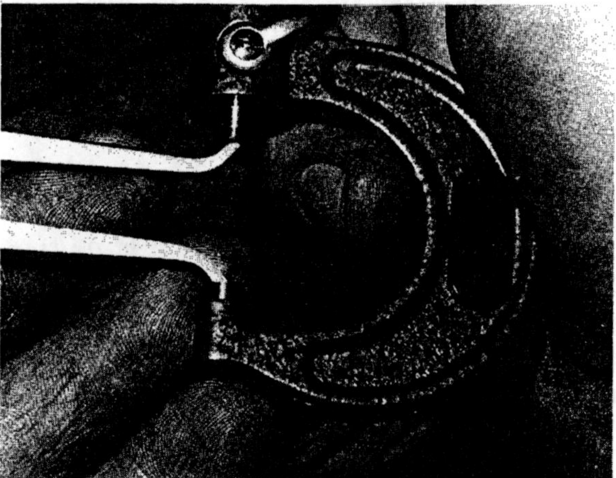

Fig. 6-5 Inside caliper measured and directly read with a micrometer after being set to the workpiece hole

CLASSIFICATION OF MEASURING TOOLS

Measuring tools may be classified in one of two ways, depending on the degree of measuring accuracy. *Semiprecision* measuring tools include those tools which easily measure to within 1/64 of an inch (0.4 mm), such as tapes and rules. *Precision* measuring tools are those which are capable of measuring more exact dimensions less than 1/64 of an inch (0.4 mm). These tools include micrometers, Vernier calipers, and dial indicators. These tools are discussed in Unit 7.

Other tools, such as straightedges, are used with linear measuring tools to determine flatness and straightness of work. Roundwork measurements are usually made by contact, using tools with contact points or surfaces. These include spring calipers, micrometers, Vernier calipers, and dial gages.

CONTACT MEASUREMENTS

Contact measurements are made in two ways. The first method is by pre-setting the tool to the required dimension using other tools as a standard. The set dimension is then compared with the actual size of the work, Figure 6-6. This method is used to make repeated measurements.

The reverse of this is the second method. The contact points of the measuring tool are set to the work surfaces and directly read from a scale or micrometer. This method is used to obtain the actual size of the part.

RULES AND TAPES

Steel Rule

The *steel rule* is a basic measuring tool from which many other tools have been developed, Figure 6-7. Steel rules, or *scales* as they are often called, are among the most used tools in the shop. Rules are made in a variety of styles. They range in size from 1/4 inch to 12 feet in length.

Steel rules are *graduated*. This means the rules are marked with fine lines on each

edge of both sides and often at the ends in subdivisions of the inch and in metric measurement. Some rules provide both English and metric systems on a single rule, Figure 6-8. English system graduations are com- as fine as 0.001 inch in decimals or 1/64 inch in fractions. Metric graduations are usually as fine as 1/2 millimetre. Figure 6-9 shows how to use the rule.

Hook Rule

The *hook rule*, which is available on many rules, is very convenient, Figure 6-10. The hook rule is used for taking measurements quicker with greater accuracy by placing the

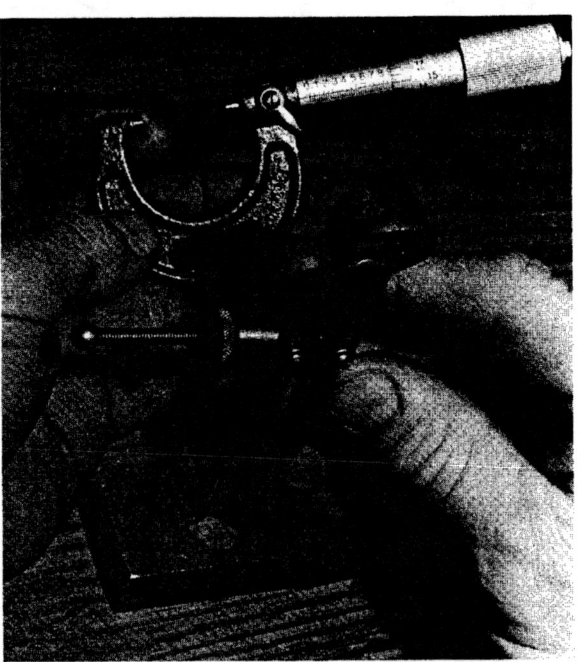

Fig. 6-6 Measuring hole with inside caliper after setting to the micrometer dimension

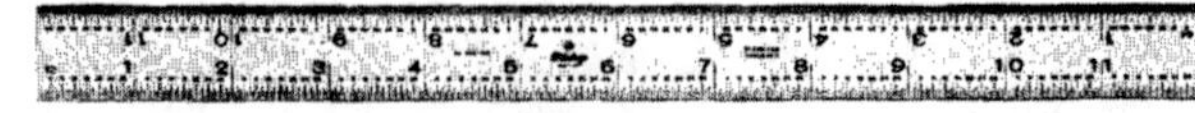

Fig. 6-7 A steel rule (MTI Corporation)

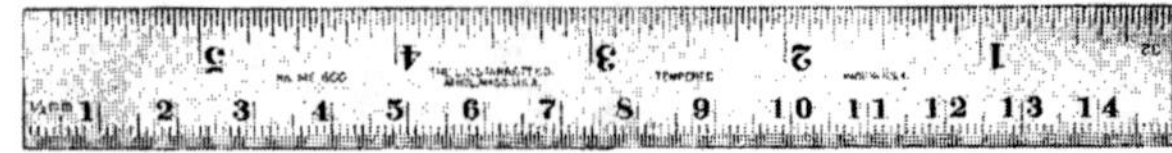

Fig. 6-8 Steel rule graduated in half millimetres and thirty-seconds of an inch (L. S. Starrett Company)

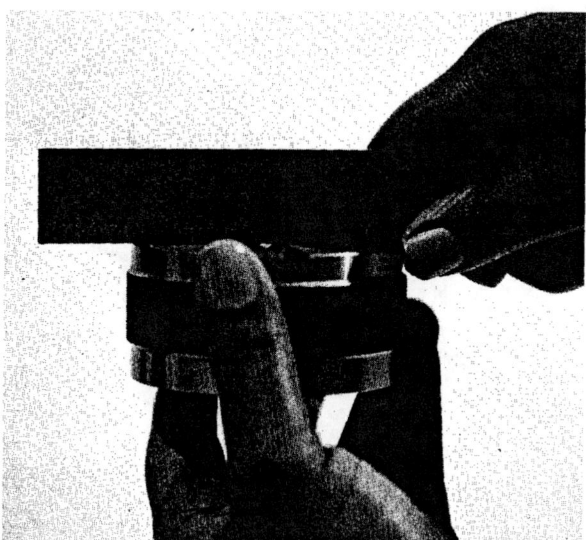

Fig. 6-9 Use of the steel rule to measure the size of a part

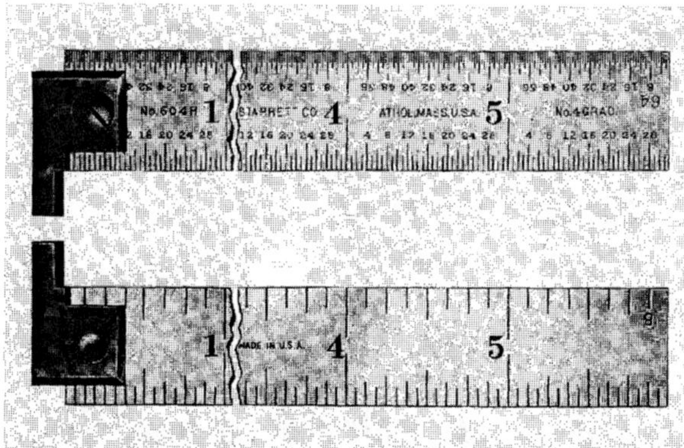

Fig. 6-10 Hook rule (L. S. Starrett Company)

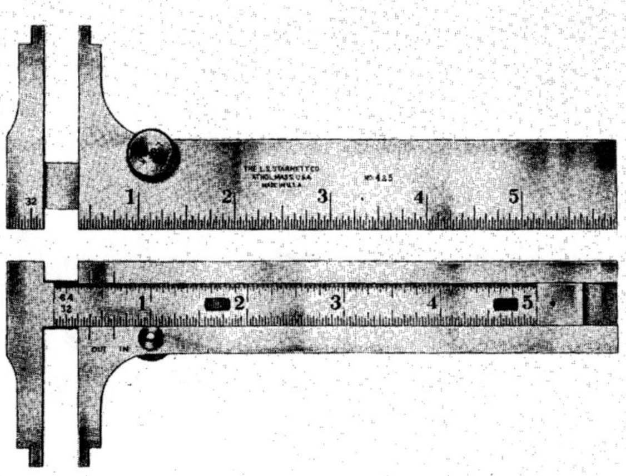

Fig. 6-11 Pocket slide caliper rule
(L. S. Starrett Company)

hook on the edge of the work. It also provides an accurate stop at the end of the rule for setting calipers.

Pocket Slide Caliper Rule

The *pocket slide caliper rule* is used to measure round bars, tubing, and other objects, Figure 6-11. The slide caliper has a pair of jaws added to the rule. One jaw is fixed at the end, and the other is movable along the scale. The movable jaw is adjusted to the workpiece and locked by a thumb screw. The slide is graduated to provide a direct reading of the inside or outside measurement of a part.

Depth Rule Gage

The *depth rule gage*, Figure 6-12, is a special rule used for measuring the depth of holes or recesses. It has a sliding base or head which is set at right angles to the rule. The base is held firmly against the surface of the workpiece and the rule is adjusted to the depth of the hole or slot. A thumb screw

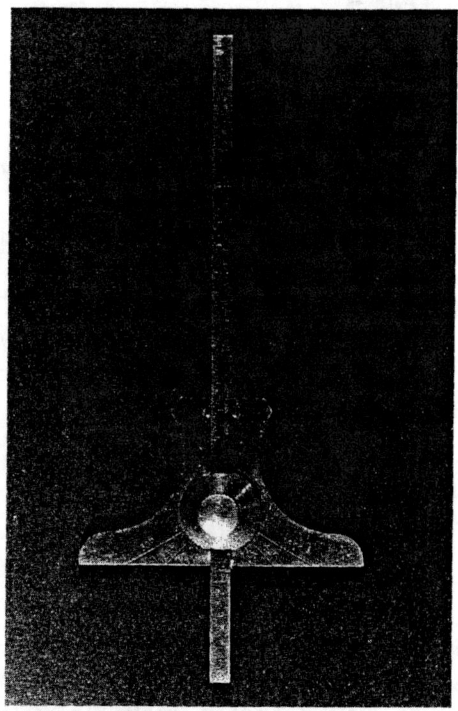

Fig. 6-12 Depth rule gage (L. S. Starrett Company)

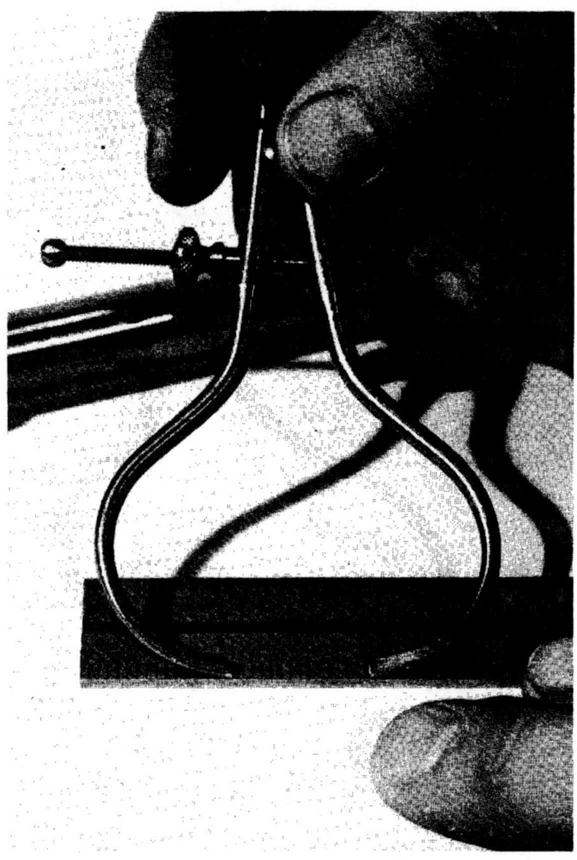

Fig. 6-13 Outside measurement being set or read on the steel rule

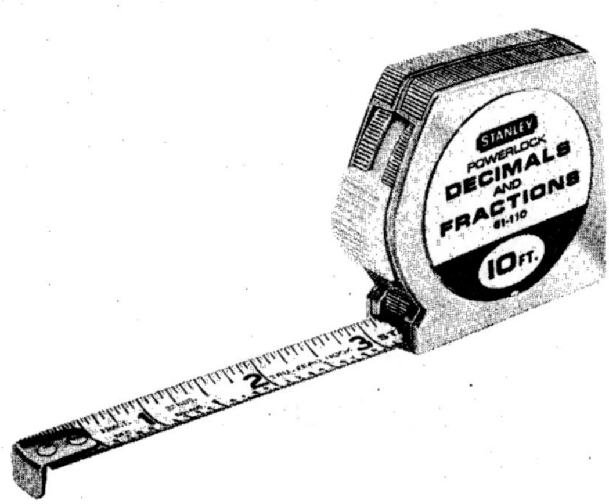

Fig. 6-14 Steel tape (The Stanley Works)

clamps the rule to the base and a direct depth reading can be taken from the rule. Other tools, commonly used in the shop for measuring include the circumference, keyseat, narrow, short, and flexible rules.

As rules are used they become worn on the ends. This results in the end inch measurement being less than one inch. To measure accurately, a good practice is to use the 1-inch mark on the rule as the reference or beginning point, Figure 6-13.

Steel Tapes

Steel tapes provide a graduated measuring tool beyond the practical limits of a steel rule, Figure 6-14. They are accurate and available in lengths up to 100 feet. Like steel rules, steel tapes are available in various graduations, including English, metric, metric and

English, and in special graduations of tenths and hundredths of a foot.

How to Read English Steel Rules and Tapes

Always reduce a fraction to the lowest terms. The fractions 2/16 inch or 4/8 inch are not used. These fractions are reduced to read 1/8 inch or 1/2 inch.

Decimal Rules

Decimal rules, Figure 6-15A, are divided or graduated in 10 parts or a multiple of 10 parts of an inch. Most decimal rules are graduated in 50 or 100 parts of an inch.

On the 50th scale in Figure 6-16, each inch is divided into 50 equal parts, with each part equal to 1/50 inch or 0.020 inch (twenty-thousandths of an inch). The scale is also marked at each 1/10-inch increment for easier reading (1/10 inch = 100 thousandths or 0.100 of an inch).

On the 100th scale, each part is divided into 100 equal parts, with each part equal to 1/100 inch or 0.010 inch (ten-thousandths of an inch). The scale is also marked at each 1/10 inch for easier reading, Figure 6-16.

Metric Rules

The typical metric rules are graduated in millimetres (mm) and half millimetres, Figure

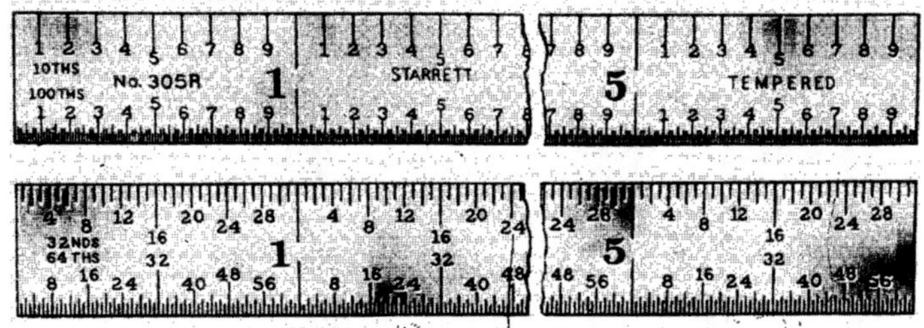

Fig. 6-15A Decimal rules (L. S. Starrett Company)

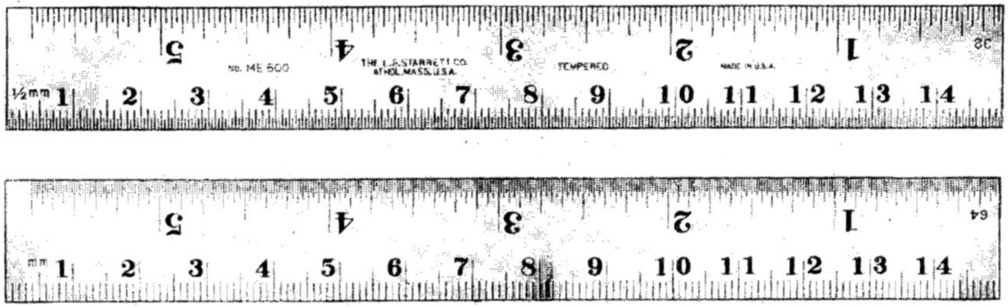

Fig. 6-15B Metric rules (L. S. Starrett Company)

6-15B. On one scale, the centimetre is divided into 10 equal parts (10 millimetres = 1 centimetre), Figure 6-17. Numbered lines are centimetre markings (10 millimetres = 1 centimetre).

The other scale represents centimetres, divided into 20 equal divisions, with each division equal to a half millimetre (0.5 millimetre), (20 half millimetres = 1 centimetre).

Care of the Rule

- Do not use a rule as a screwdriver.
- Do not use a rule to clean metal chips from a machine or the bench.
- Do not lay other tools on the rule.
- Polish the rule with fine steel wool and wipe it with a clean, soft, oily cloth to protect the accurate graduations.
- Take measurements and tool settings from the 1-inch mark rather than from the end of the rule.
- Do not measure moving parts with a rule.
- Store the rule separately and protect it.
- Use the correct rule for a job.

CALIPERS

Calipers are tools used to measure diameters and lengths. They range in size from 2 inches to 48 inches. *Bow* calipers are often

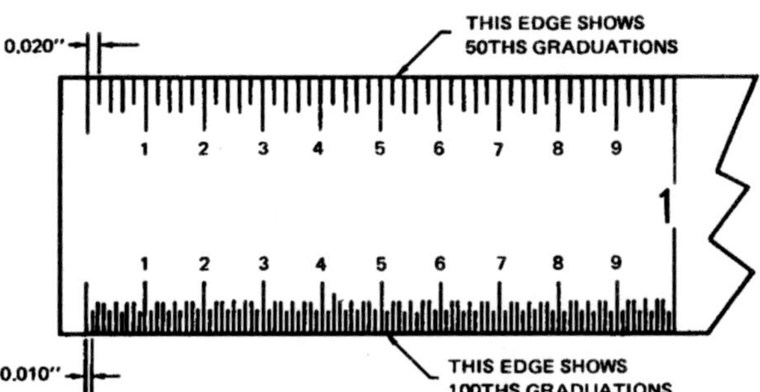

Fig. 6-16 Graduations of decimal parts of an inch

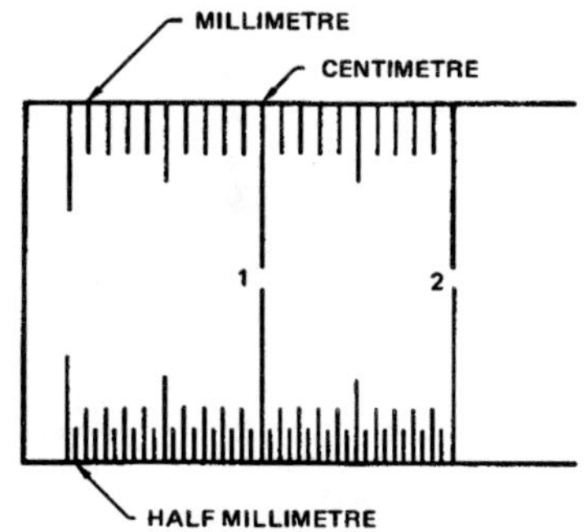

Fig. 6-17 Centimetre, millimetre, and a half millimetre shown on a metric rule

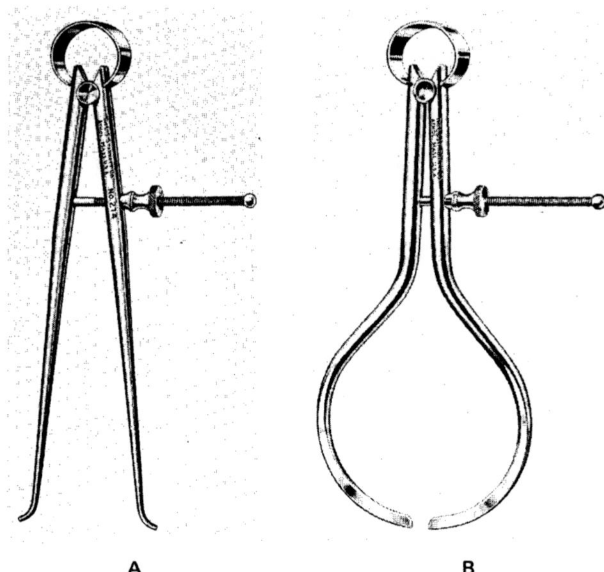

Fig. 6-18 Standard spring joint: (A) Inside caliper, and (B) outside caliper (L. S. Starrett Company)

used for contact measurement. They are useful for measuring distances between or over surfaces. They are also used for comparing dimensions or sizes with standards such as graduated rules.

Caliper Styles

Calipers are made in several styles, such as spring-joint, firm-joint, and lock-joint, Figure 6-18. Calipers with the legs shaped either for inside or outside measurements are made in all three styles.

The spring bow-style calipers use an adjusting nut and screw working against the tension of the top spring clamp. The firm-joint style is made so the tension of a nut and stud provides enough friction to hold the legs in any set position. The lock-joint style is made

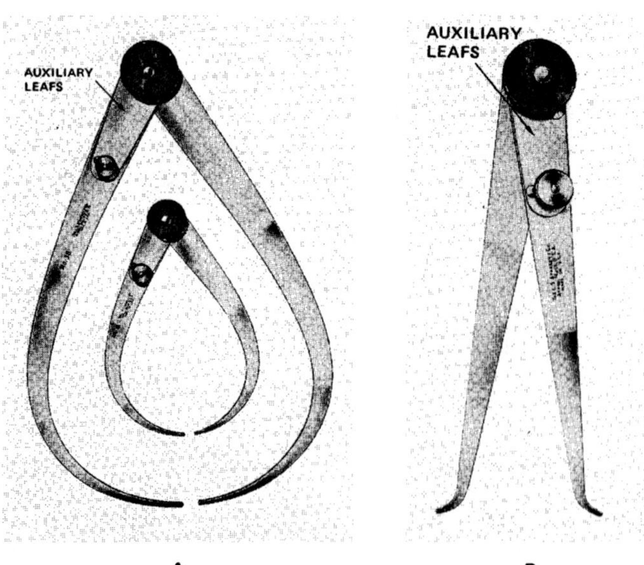

Fig. 6-19 Transfer calipers:
(A) Outside lock-joint caliper, and (B) inside lock-joint caliper (L. S. Starrett Company)

with a knurled nut which may be loosened for free movement of the legs and tightened to lock the setting.

Transfer calipers are of the lock-joint style made with a stud or stop on a freely moving leg fitting into a slot or against a stop on an auxiliary leaf, Figure 6-19A and B. The free leg can be moved in or out to clear collars, flanges, or other obstructions. It can then be returned to the original setting against the stop to make the measurement.

How to Use Calipers

The working tips of the caliper legs should always be kept in line to maintain the tool's accuracy, Figure 6-20. The accurate use of calipers depends on the individual's feel or sense of touch. This is developed through practice and use. Calipers must not be forced over external work or into internal work. The caliper legs should always be square with the axis of the workpiece being measured.

Use the following steps as a guide in the proper use of the outside caliper.

1. The caliper must be held by hand in a firm, delicate manner with the legs in contact with the work, Figure 6-21.
2. Adjust the knurled adjusting nut with the thumb and forefinger until the weight of the caliper leg causes it to slip over the work.
3. Carefully remove the caliper from the work without disturbing the setting.
4. Measure the distance between the caliper legs with a steel rule and read the dimension, Figure 6-22.

Note: The caliper may be used to check a given size by first setting it to a desired dimension from a steel rule or other direct reading tool. The caliper is then used as a gage to machine a part to the required size. The steps listed below are a guide to correctly set and use inside calipers.

1. The caliper is again held by hand in the same firm, delicate manner. Have one

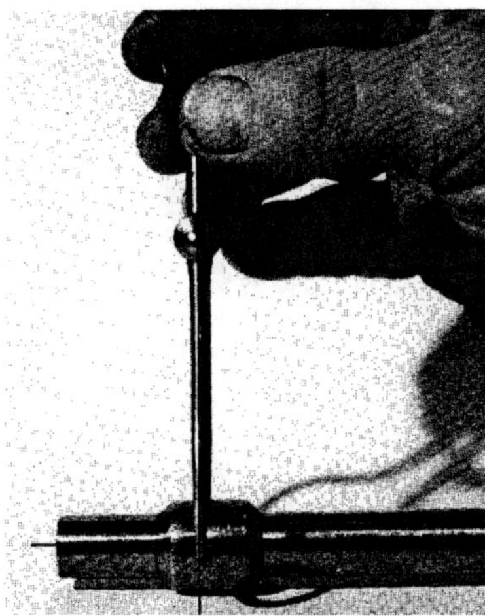

Fig. 6-20 Caliper tips must be kept in line to measure accurately.

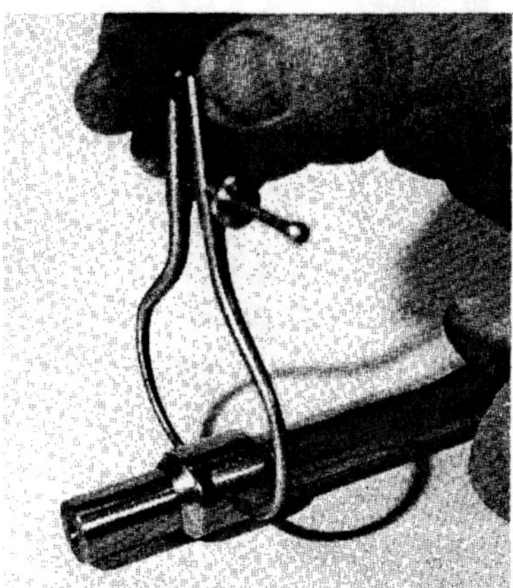

Fig. 6-21 Correct method to hold the caliper to the work for accurate measurement

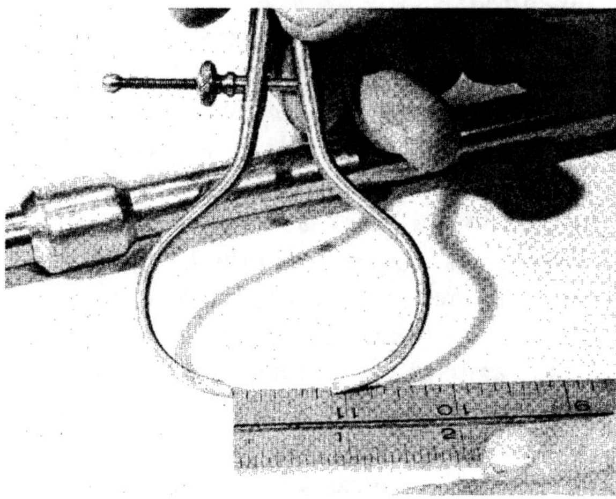

Fig. 6-22 Measure the distance between the caliper with a steel rule.

Fig. 6-23 Hold inside calipers square and across the center of the workpiece for accurate measurement.

Fig. 6-24 Arc the caliper leg as shown by the arrow while holding A stationary to the work.

leg resting slightly inside of the space to be measured, Figure 6-23.

2. The adjusting nut is turned to bring the other leg gently in contact with the opposite side of the space or hole being measured.

3. Arc the caliper leg by pivoting on the rested leg. Continue adjusting the caliper until the correct feel or drag of the arc leg is felt inside the space, Figure 6-24.

4. Make sure no side motion exists in the tips of the caliper legs. Also, make sure these tips are square with the largest part of the diameter being measured.

5. When measuring the distance over the inside caliper legs, use a steel rule. Place one end of the rule against a solid, true surface. One leg of the caliper is rested against this surface, and the dimension is read from the scale graduations, Figure 6-25.

Fig. 6-25 Read the measurement of the distance on a steel rule.

Note: When reading the scale, always look squarely at the rule and have the light fall directly on the rule. When setting the caliper, be sure the caliper tip splits the line of the graduation. Make sure that the ends of the rule are not worn or damaged when setting to or measuring with a rule.

6. The inside caliper is also used to check a given size by first setting it to the desired dimension, and then using the caliper as a gage to check the inside diameter.

Calipers are not efficient tools to use for accurate measurements. Calipers are generally used for rougher machining operations, measuring stock, lengths, and measuring or gaging dimensions that need not be extremely accurate. The micrometer is a much quicker and more dependable method to obtain precision measurements. This tool is discussed in Unit 7.

GAGES FOR MEASURING

Telescope gages, Figure 6-26, are tools used to transfer measurements of internal dimensions. To use the gage, it is inserted into a hole and allowed to expand from the internal spring tension against the plungers. After the proper feel is obtained, the contact plungers are locked by a slight turn of the knurled screw in the handle.

Telescope gages should be side wiggled in the hole to be sure the greatest diameter is measured. One contact rests on the side of the hole. The other contact pivots about this contact through the center of the hole in a similar manner to the inside caliper.

The size of the hole is determined by removing the gage from the hole and measuring the distance across the two contacts with a micrometer caliper.

Small hole gages are made to accurately gage the dimensions of smaller holes, Figure 6-27. The gage is used in the same manner as the telescope gage and adjusted to size by a knurled knob at the end of the gage. The gage is removed from the hole and measured to determine the size with an outside micrometer. Sizes range from 0.125 inch to 0.500 inch.

MISCELLANEOUS MEASURING TOOLS AND GAGES

Several miscellaneous tools and gages are used to measure at the bench. The more commonly used are described below.

Drill point gages are used to accurately measure the length and angle of the cutting lips of twist drills, Figure 6-28. *Drill-size and wire-size gages* are used to quickly and conveniently

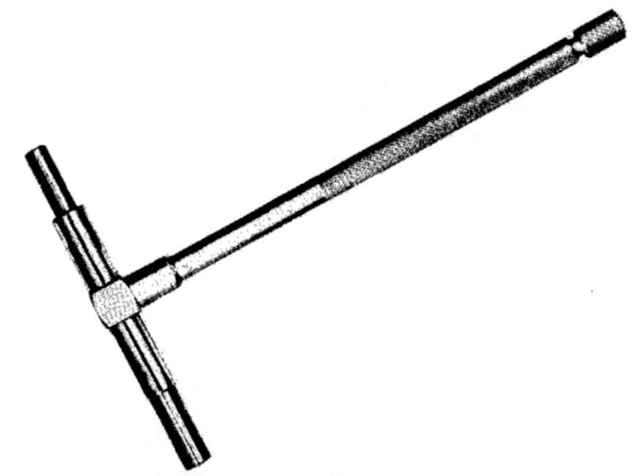

Fig. 6-26 Telescope gage (MTI Corporation)

Fig. 6-27 Small hole gage (MTI Corporation)

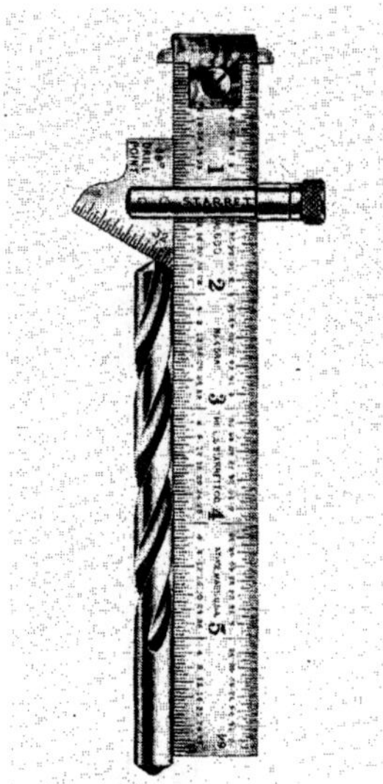

Fig. 6-28 Drill point gage (L. S. Starrett Company)

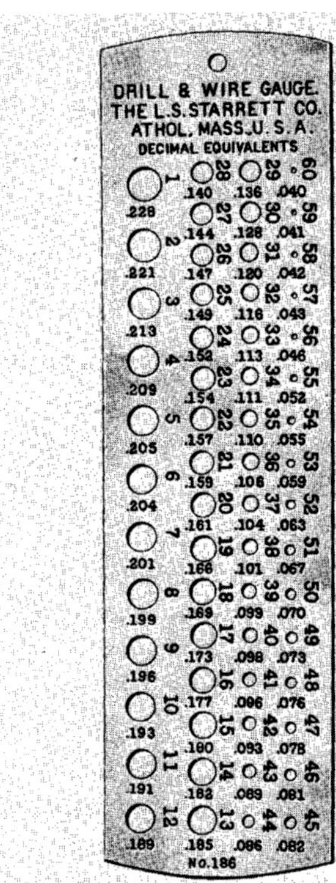

Fig. 6-29 Drill-size and wire-size gage
(L. S. Starrett Company)

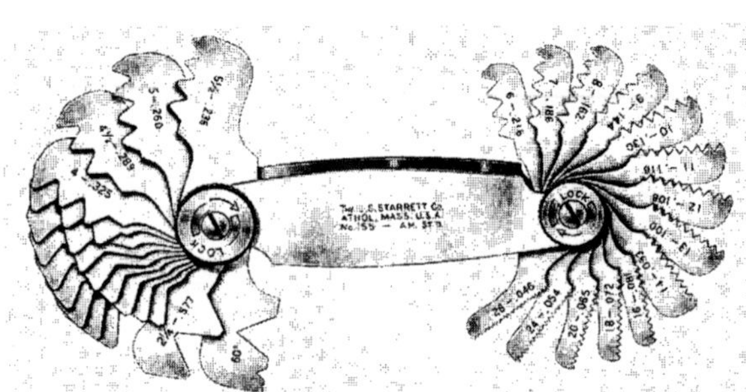

Fig. 6-30 Screw pitch gage (L. S. Starrett Company)

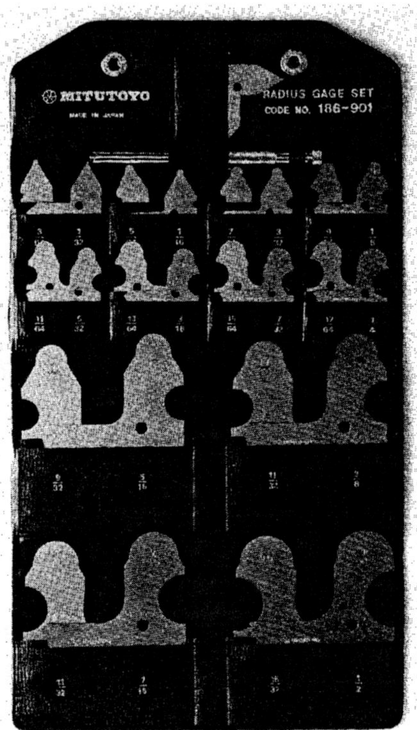

Fig. 6-31 Radius gages (MTI Corporation)

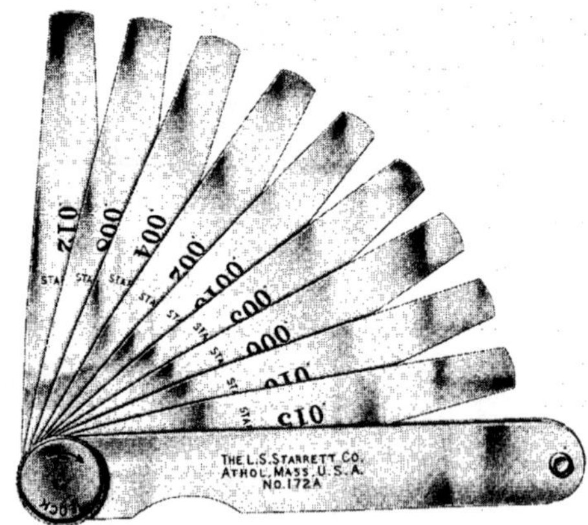

Fig. 6-32 Feeler gage (L. S. Starrett Company)

check the size of twist drills and steel drill rod. Figure 6-29 shows a combined drill and wire-size gage.

Screw pitch gages are made to match the teeth of one of the leaves with the thread on the work, Figure 6-30. The correct pitch of the thread can be read quickly from the leaf. These gages are made for English and metric threads. *Radius gages* are used to check or layout radius or fillets ranging from 1/64 inch to 1/2 inch, Figure 6-31.

Thickness feeler gages are made to check the clearances between parts, Figure 6-32. Individual leaves are marked with the thickness in either English or metric sizes.

Many other special measuring tools are used by the worker to complete required work at the bench. These tools are available and presented in the many tool catalogs from suppliers.

REVIEW QUESTIONS

A. Multiple Choice

1. Which of the following is used to measure round bars and tubing?
    a. Depth rule gage
    b. Hook rule
    c. Pocket slide caliper rule
    d. Steel rule

2. Which of the following gages is used to transfer measurements of internal dimensions?
    a. Telescope gage
    b. Hole gage
    c. Feeler gage
    d. Radius gage

3. Which of the following is the smallest fractional part of an inch that the English rule is divided into?
    a. 1/8 inch
    b. 1/32 inch
    c. 1/64 inch
    d. 1/16 inch

4. Which of the following identifies the finest graduation applied to metric rules?
    a. 1 millimetre
    b. 1/2 millimetre
    c. 1 centimetre
    d. 0.02 millimetre

5. Which of the following is not a direct-reading measuring tool?
    a. Steel rule
    b. Telescoping gage
    c. Micrometer
    d. Dial indicator

6. Steel rules may be cleaned by using which of the following?
    a. Smooth file
    b. Fine emery paper and oil
    c. Fine sandpaper and oil
    d. Fine steel wool and oil

B. Short Answer

7. List the two methods in which linear measurements may be made.

8. List the three different types of readings available on steel rules.

9. State the three joint styles of calipers.

10. Explain the term *graduated*.

11. _____________ calipers are useful for measuring distances between or over surfaces.

ACTIVITIES

1. Using calipers, transfer measurements from finished parts and read the dimensions from a steel rule.

2. From a print, set the caliper with a rule and check the finished part size.

UNIT 7 PRECISION MEASURING TOOLS AND GAGES

OBJECTIVES

After completing this unit, the student will be able to

- define precision measuring responsibility, reliability, and discrimination.

- identify the common precision measuring tools used in the shop.

- accurately set and read precision measuring tools, including metric tools.

INTRODUCTION

New measuring tools are continually being developed to meet the expanding precision needs of industry. Easier, quicker reading tools are being made to remove the chances of human error in measuring and reading the measured dimension. Some of these tools have digital readout with the exact dimension displayed. Other tools are using light beams to measure very accurately.

USING PRECISION TOOLS

Learning to use precision measuring tools requires practice of good measuring habits. To develop good habits of consistent, accurate measurement, the worker relies upon the sense of sight and touch. The sense of touch is important when using contact measuring tools. This sense of touch (feel) is most prominent in the fingertips. Therefore, a contact measuring tool should be properly balanced in the hand and held lightly to allow the fingers to move the tool. Holding and using the tool in this manner greatly enhances the sense of touch to obtain accurate measurement. The sense of sight needs little explanation. Without it, setting and reading the measurement is very difficult.

MEASURING RESPONSIBILITY

The *responsibility* of the benchworker is to know that the measuring tool is properly adjusted. The benchworker is also responsible for determining from the print the degree of accuracy the measurement must meet. Other responsibilities include the degree of measuring-tool discrimination and which tool will be the most reliable for the job.

The worker must also be very careful to position the measuring tool properly to obtain an accurate measurement. Accurate measurements must be made with the measuring tool exactly in line with the axis of the measurement. The accurate, in-line measurement usually defines the size of the part.

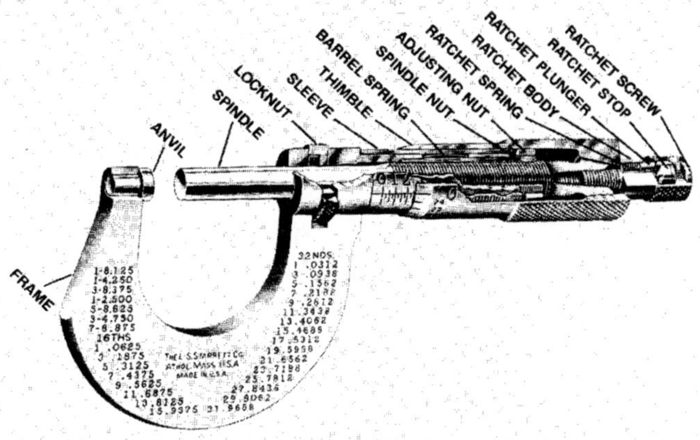

Fig. 7-1 A micrometer, or mike (L. S. Starrett Company)

RELIABILITY AND DISCRIMINATION OF MEASURING TOOLS

A measuring tool is considered reliable when it is not used beyond its discrimination. *Discrimination* refers to the degree to which a measuring tool divides the basic unit of length that it is using for measurement. *Reliability* in measuring refers to the ability of a measuring tool to obtain readings to the degree of accuracy required. Make sure the proper measuring tool is used to measure to the required degree of accuracy. For example, a steel rule should not be used to measure thousandths of an inch.

A micrometer can subdivide an inch into 10,000 equal parts. Some micrometers can discriminate to 0.0001 of an inch, but they are most accurate measuring to an accuracy of 0.001 of an inch. The 0.001-inch measurement would be 1/10 of the finest measurement possible on this micrometer. An example of this ratio would be a micrometer, Figure 7-1.

Micrometers must be set to zero with a known standard, Figure 7-2. The worker must make sure the measuring tool is accurately adjusted. Any inaccuracy of the tool causes the same error in measurement of parts.

MICROMETER CALIPERS

The micrometer caliper, often called a mike, offers the benchworker a much more versatile and economical way to measure. A worker can easily and quickly obtain and directly read a measurement with the micrometer caliper. Identical measurement can be made again and again with the use of friction and ratchet devices to preset the correct feel.

The micrometer is an important measuring tool because of its convenient size, strength, accuracy, easy adjustment, and readability. Micrometer calipers are made in standard sizes from 1/2 inch to 60 inches.

Parts of the Micrometer

A sectional view of the standard micrometer caliper is shown in Figure 7-1. The six major parts are the frame, anvil, spindle,

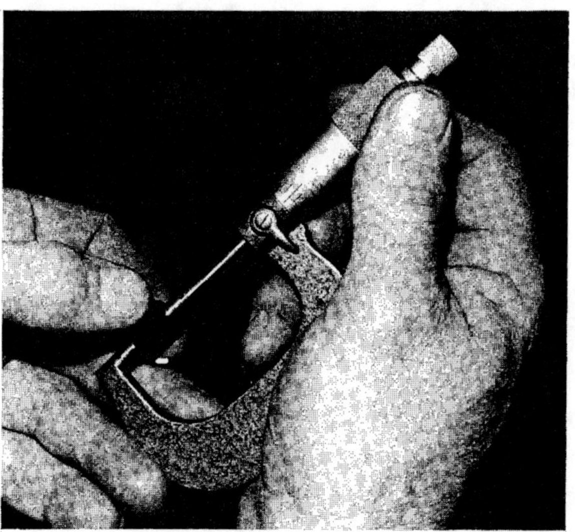

Fig. 7-2 Checking micrometer with a known standard

locknut, sleeve, and thimble. Ratchets and friction thimbles are optional.

As the above figure shows, the screw threads on the spindle rotate inside a fixed nut hidden by a sleeve. As the screw rotates, the opening between the anvil and the spindle faces varies. The graduations on the sleeve and thimble indicate precisely the screw position and the amount of the opening between the faces. The thimble rotates with the screw spindle and travels along the barrel. Each 0.025 inch graduation on the sleeve matches the pitch, 1/40 inch, of the spindle measuring screw, one line for each revolution. The distance between the spindle and anvil faces changes by 1/40 inch as each line is exposed on the sleeve.

Reading the Micrometer

On the micrometer, the sleeve is marked with 40 lines. Each line represents 0.025 of an inch, Figure 7-3. Every 4th line is drawn longer and numbered to indicate 0.1 of an inch, 0.2 of an inch, etc.

The bevel edge of the thimble is divided into 25 equal parts. Each of these parts represents 1/25 of the distance the thimble travels along the barrel in moving from one 0.025 division to the next. Thus, each division on the thimble represents 0.001 of an inch. These divisions are marked at every 5 spaces by 0, 5, 10, 15, and 20. When 25 of these graduations have passed the horizontal line on the barrel, the spindle has moved 0.025 of an inch.

With all micrometers, always be sure to add the basic micrometer size (1″, 2″, 3″, etc.) into the measurement.

Example: The micrometer shown in Figure 7-4 is read as follows: Note the basic size "0" and the last visible numbered line on the sleeve. It is 0.100 inch. There are 3 more shorter lines past the numbered lines. Each line represents 0.025 inch. Therefore, 3 × 0.025 inch equals 0.075 inch. The divisions on the bevel edge of the thimble each represent 0.001 inch. There are 3 divisions beyond the 0 mark. Therefore, the total is 0.003 inch. To obtain the total, add the 3 numbers found in each of the steps above.

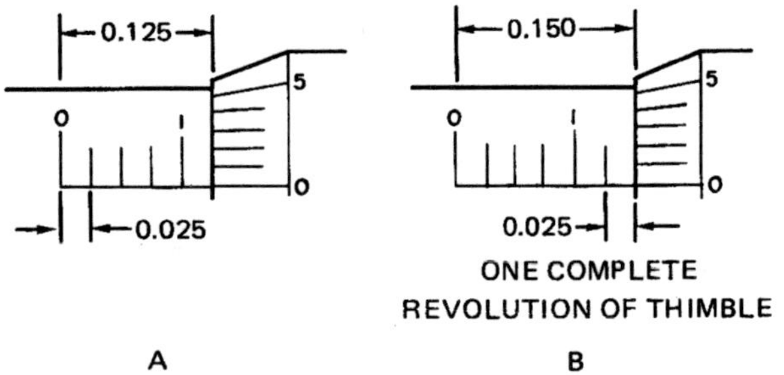

Fig. 7-3 Divisions of 0.025 inch

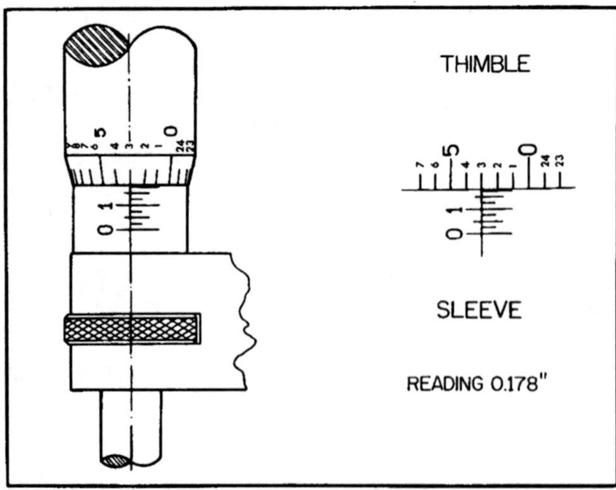

Fig. 7-4 Reading a micrometer (L. S. Starrett Company)

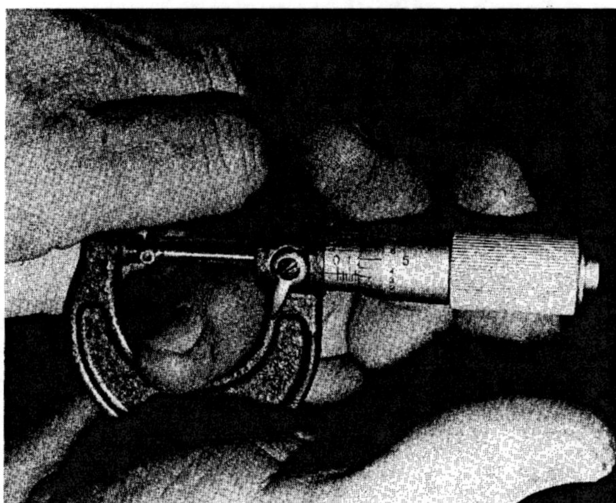

Fig. 7-5 Proper use of mike to measure 0.178-inch diameter part

Basic micrometer size	0.000 inch
Tenths on sleeve	0.100 inch
Visible sleeve graduations past the last tenth.	0.075 inch
Thimble division aligning with the sleeve index.	0.003 inch
Total.	0.178 inch

This reading represents the opening between the anvil and spindle measuring surfaces of the micrometer. Figure 7-5 also shows proper use of the mike to obtain accurate measurements.

VERNIER MICROMETER

The Vernier micrometer has a Vernier scale wrapped around the sleeve, Figure 7-6. This is the only difference between it and the standard micrometer.

This scale has 10 divisions marked on the sleeve which occupy the same space as 9 divisions on the beveled edge of the thimble. The difference, between the width of one of the 10 spaces on the sleeve and one of the 9 spaces on the thimble, is 1/10 of a division on the thimble. Since the thimble is graduated to read in thousandths, 1/10 of each thimble division is 0.0001 of an inch.

To make the reading, read to 0.001 of an inch as with the standard micrometer. Next, add the horizontal, Vernier scale line on the

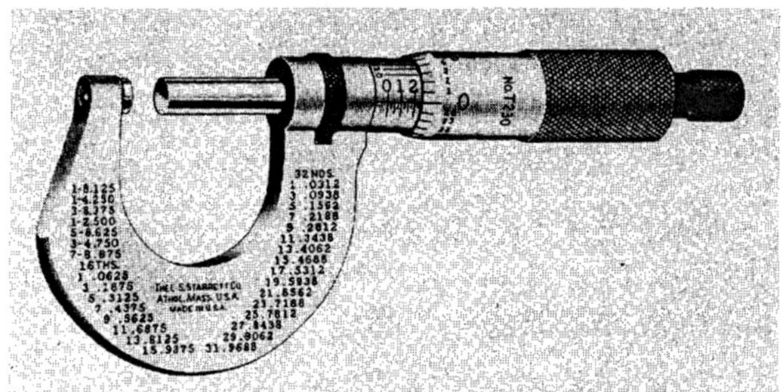

Fig. 7-6 Vernier micrometer (L. S. Starrett Company)

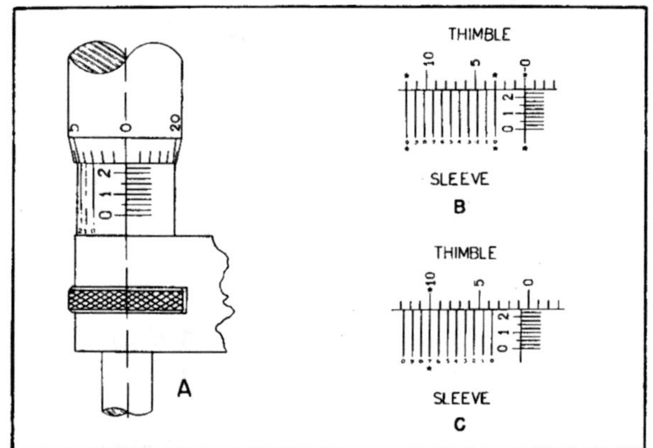

Fig. 7-7 Reading a Vernier micrometer (L. S. Starrett Company)

sleeve that exactly coincides with a line on the thimble. These lines are numbered 0 to 0, meaning from 0 to 10.

Note: Read only the number on the horizontal line of the Vernier scale which exactly lines up with a line on the thimble. Do not read the numbers or count the lines on the thimble.

Reading the Vernier Micrometer

Example: In Figure 7-7A, the zero on the thimble coincides exactly with the axial line on the sleeve. The Vernier 0 on the sleeve coincides with a line on the thimble. Therefore, the reading is an even 0.2500 inch. In Figure 7-7B, the 0 line on the thimble has gone beyond the axial line on the sleeve, indicating a reading of more than 0.2500 inch. Checking the Vernier scale in Figure 7-7C shows that the 7th Vernier line on the sleeve is the one which exactly coincides with a line on the thimble. Therefore, the reading is:

0.2500 inch + 0.0007 inch = 0.2507 inch

METRIC MICROMETER

In the metric system, the familiar 1-inch micrometer becomes a 25-millimetre micrometer. The tools look alike, with the only difference being the graduations and the pitch of the threads on the spindle.

Since the pitch of the spindle screw is 0.5 mm, one revolution of the thimble opens or closes the space between the anvil and spindle the same 0.5-mm distance. The reading axis line on the sleeve is graduated in 1.0 mm. Every 5th millimetre is numbered from 0 to 25. Each millimetre is also divided in 0.5 mm, and 2 revolutions of the thimble are required to advance the spindle 1.0 mm.

The beveled edge of the thimble is graduated in 50 divisions. Every 5th line is numbered from 0 to 50 since one revolution of the thimble advances or withdraws the spindle

0.5 mm, each thimble graduation equals 1/50 or 0.5 mm or 0.01 mm. Thus, 2 thimble graduations equal 0.02 mm, 3 graduations equal 0.03 mm, etc.

Reading The Metric Micrometer

To read the micrometer, add the number of millimetres and half millimetres visible on the sleeve to the number of hundredths of a millimetre indicated by the thimble graduation which coincides with the reading line on the sleeve.

Example: In Figure 7-8,

The 0.5-mm sleeve graduation
is . . . .   5.00 mm
One additional 0.5-mm line is
visible on the sleeve. This
equals . . . .   0.50 mm
Line 28 on the thimble coincides with the reading line
on the sleeve. Thus, 28 ×
0.01 mm equals. . . . .   0.28 mm
To get the total reading,
add the 3 numbers found
above.   5.78 mm

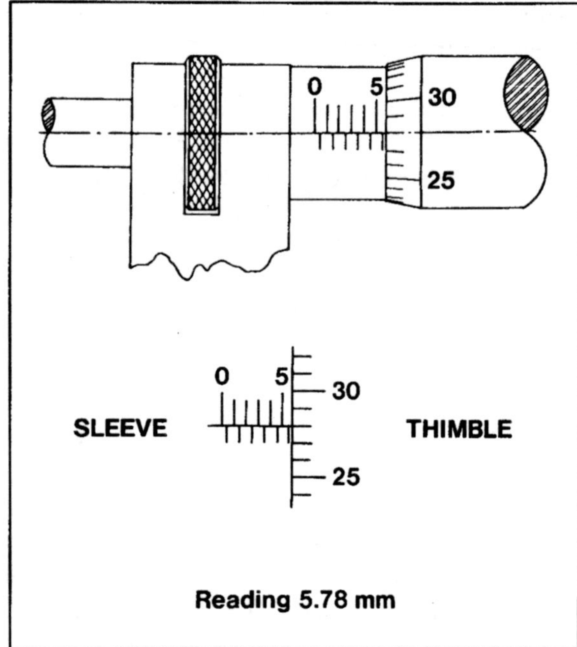

Fig. 7-8 Reading the metric micrometer (L.S. Starrett Company)

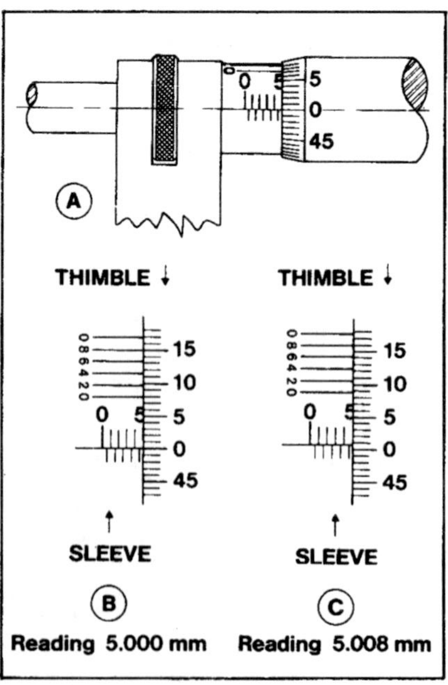

Fig. 7-9 Reading the metric Vernier micrometer (L.S. Starrett Company)

Reading The Metric Vernier Micrometer

Obtain a reading to 0.01 mm in the same way shown in Figure 7-9 for the hundredths of a millimetre. Next, find which line on the Vernier scale coincides with a line on the thimble. If it is the line marked 2, add 0.002 mm. If it is the line marked 4, add 0.004 mm etc.

Example: Referring to drawings A and B in Figure 7-9,:

The 5-mm sleeve graduation is visible.	5.000 mm
No additional lines on the sleeve are visible	0.000 mm
Line 0 on the thimble coincides with the reading axis line on the sleeve	<u>0.000 mm</u>
Therefore, the micrometer reading is.	5.000 mm

Example: Referring to drawing C in Figure 7-9,:

The 5-mm sleeve graduation is visible.	5.000 mm
No additional lines on the sleeve are visible	0.000 mm

Line 0 on the thimble lies just below the reading line on the sleeve, indicating that a Vernier reading must be added.

Line 8 on the Vernier coincides with a line on the thimble	<u>0.008 mm</u>
Therefore, the Vernier metric micrometer reading is	5.008 mm

The ability to measure very small dimensions and directly read that size with the micrometer makes it an accurate tool to work with.

HOW TO MEASURE WITH THE MICROMETER

1. Make sure the micrometer is adjusted accurately to zero or set to a known precision standard gage.
2. The anvil is held against the work, Figure 7-10.

3. The spindle is adjusted down to lightly contact the work, Figure 7-11. The mike should be held in the same firm, delicate manner as the caliper.
4. The mike is side wiggled as the spindle lightly contacts the work. This insures the anvil and spindle are seated squarely upon the work. At the same time, the entire micrometer is pivoted around the anvil. To obtain an accurate reading, the

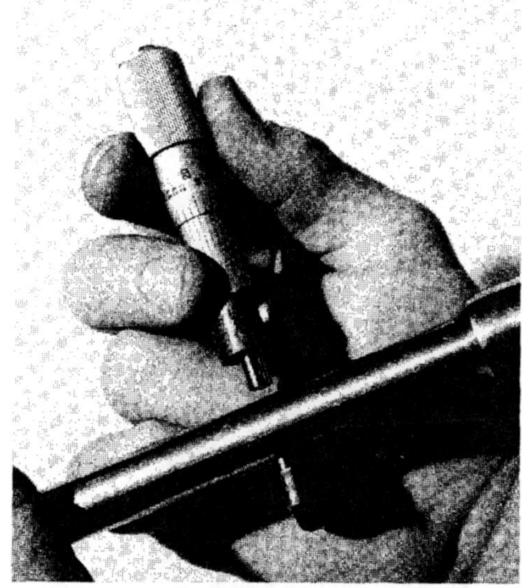

Fig. 7-10 Hold the anvil against the work.

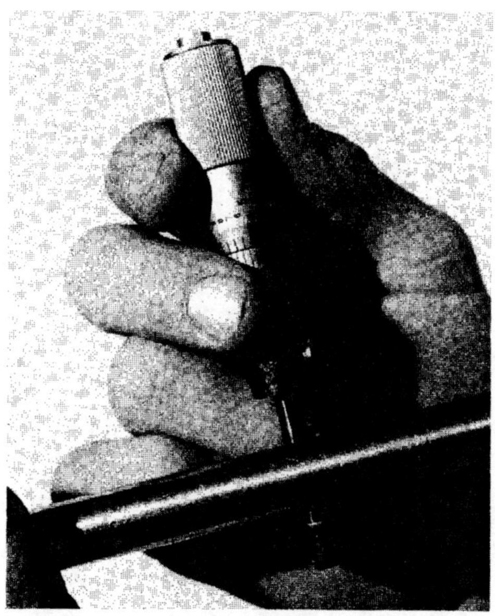

Fig. 7-11 The spindle is adjusted to the work.

anvil and spindle are eased to seat square-ly on the true diameter of the work. With the aid of the friction thimble or ratchet device on the micrometer, the thimble slips when the proper adjusting pressure is made.

5. After the measurement has been determined by the proper feel or slip of the thimble, lock the spindle locknut, remove the mike from the work, and read.

MEASURING PRECAUTIONS AND MICROMETER CARE

- Only measure work that is stationary and cooled to room temperature.
- Lubricate the moving and external parts with proper tool and instrument oil as shown in Figure 7-12.
- Never over tighten (cramp) the spindle against the work.
- Keep the anvil and spindle clean. Dirt between the anvil and spindle causes an incorrect micrometer reading.
- Wear of the screw or contact surfaces of the anvil and spindle may cause the index line and the thimble 0 line to not coincide. Readjustment is easily made. Turn the friction sleeve by means of a small spanner wrench until the 0 aligns again, Figure 7-13.
- To adjust the spindle thread wear, use the spanner wrench to tighten the adjusting nut on the split tapered spindle nut, Figure 7-14.
- Hold the mike properly and set and adjust it carefully. When measuring, the mike should be adjusted square across the diameter of the work.
- Read the mike accurately (reading over or under 0.025 inch can be a common error).
- Store the mike protected from damage with the anvil and spindle not touching.
- Use micrometers with hard-wearing carbide tips for long-lasting accuracy.
- Do not use a micrometer to measure byond its intended range.

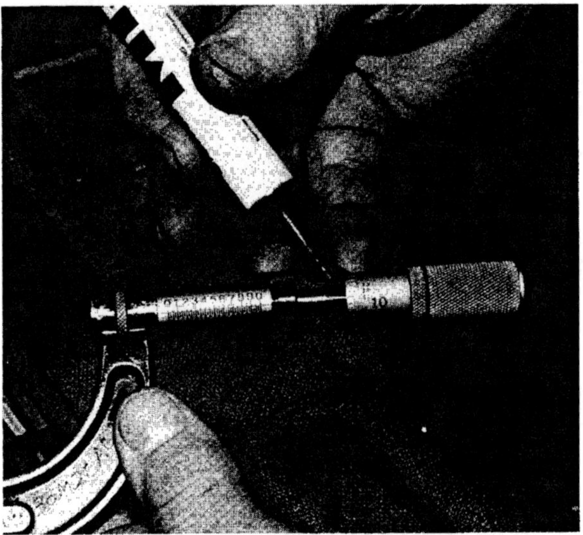

Fig. 7-12 Lubricate moving parts of measuring tools.

Fig. 7-13 Readjustment of the friction sleeve

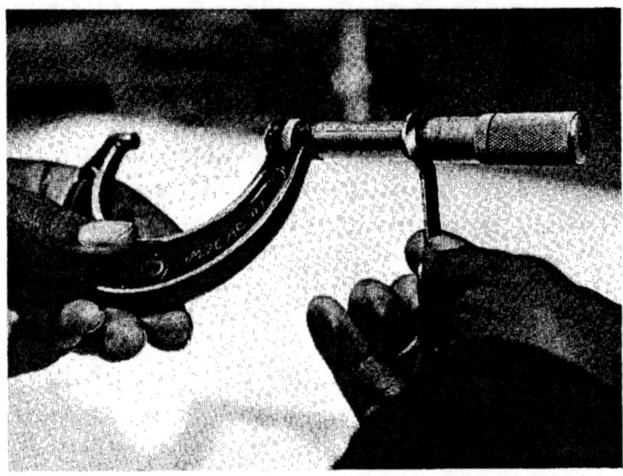

Fig. 7-14 Adjusting spindle thread wear with a spanner wrench

Fig. 7-15

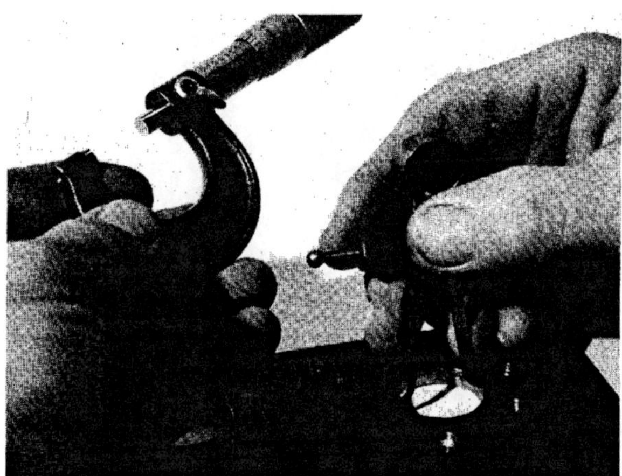

Fig. 7-16 Setting caliper to space to be measured

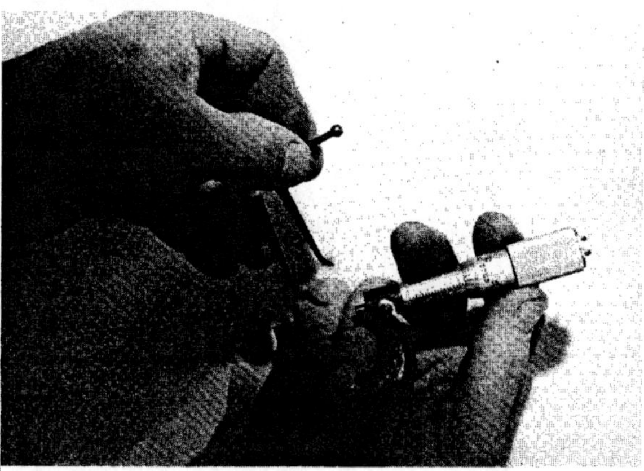

Fig. 7-17 Proper way to hold micrometer and caliper

- To open and close the micrometer a considerable distance, hold it by the frame and roll the thimble along the forearm or palm of the hand, Figure 7-15.

MEASUREMENT TRANSFER WITH THE MICROMETER

The following steps may be used to transfer inside measurements taken by an inside caliper to a direct-reading outside micrometer.

1. Set the caliper to the space being measured, Figure 7-16.
2. Hold the caliper and micrometer as shown in Figure 7-17.
3. Rotate the thimble of the micrometer until the tips of the caliper legs can be felt in light contact with the anvil and the end of the spindle.
4. Rest the tip of one leg of the caliper on the micrometer anvil and swing an arc with the other leg of the caliper. When light contact is felt at the high point of the arc, an accurate reading can be taken, Figure 7-18. The tips of the caliper legs must be held parallel to the axis of the

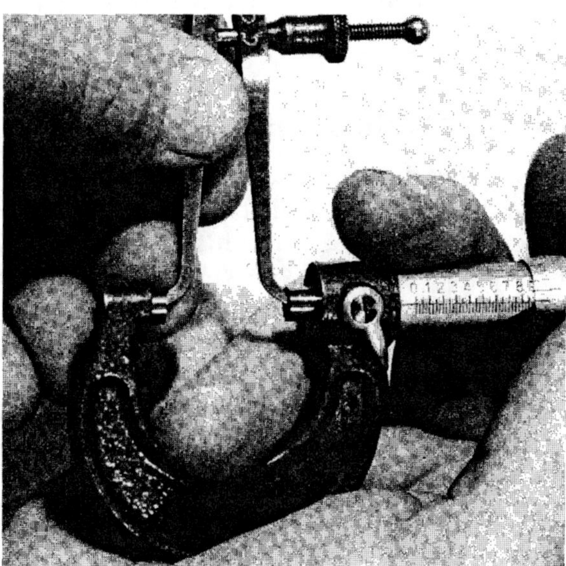

Fig. 7-18 Reading can be taken when light contact is felt between anvil and caliper leg

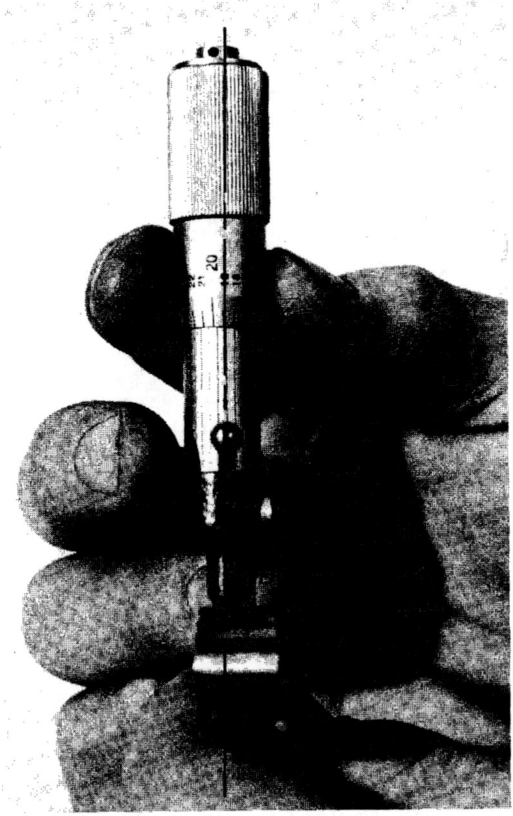

Fig. 7-19 Tips of caliper legs parallel to axis of micrometer

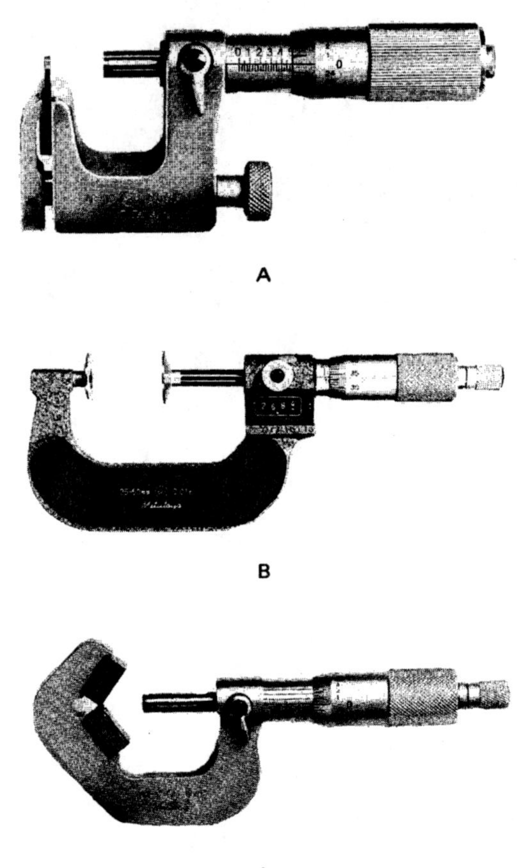

Fig. 7-20 Other styles of micrometers: (A) Multi-anvil micrometer, (B) disc-type micrometer, and (C) V-anvil micrometer (MTI Corporation)

micrometer spindle, Figure 7-19. The caliper tips pass between the anvil and spindle by their own weight when the micrometer is accurately set to them.

OTHER MICROMETERS

Many shapes, styles, and sizes of micrometer calipers are available for special purposes, Figure 7-20. Consult supplier catalogs for complete listings.

Micrometer depth gages are used to accurately measure the depth of holes, slots, and projections, Figure 7-21. Figure 7-22 shows the correct use of the tool to take a reading. The micrometer depth gage reads in the reverse of a standard micrometer.

Inside micrometers are an application of the micrometer screw principle to adjustable-end measuring gages. The distance between ends or contacts is changed by rotating the

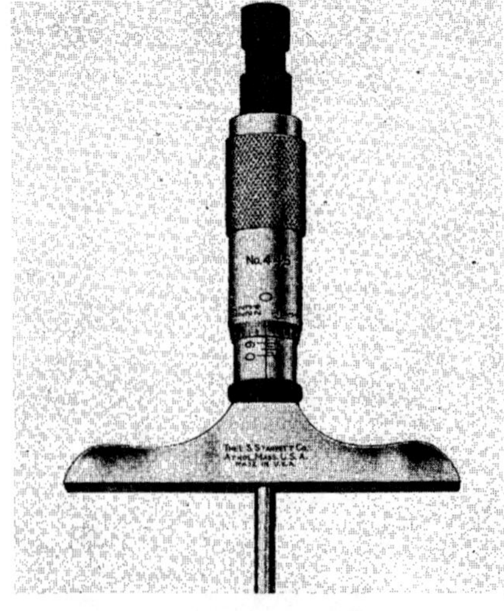

Fig. 7-21 Micrometer depth gage (L. S. Starrett Company)

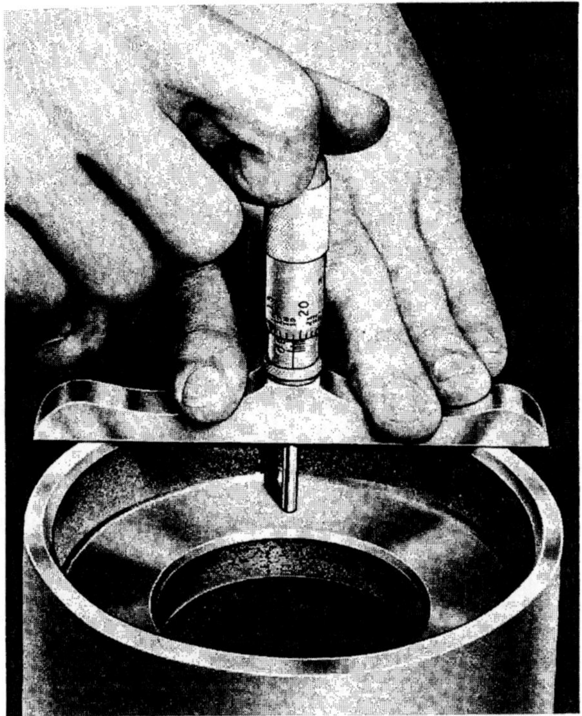

Fig. 7-22 Use of the micrometer depth gage to take a reading (L. S. Starrett Company)

Fig. 7-23

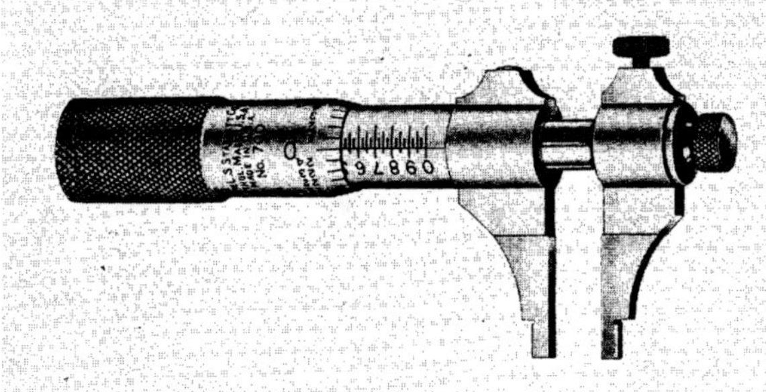

Fig. 7-24 Inside micrometer caliper (L. S. Starrett Company)

sleeve on the micrometer head to the extent of the screw length, usually either 1/2 inch or 1 inch. Greater distances are obtained by means of interchangeable extension rods and suitable collars. Figure 7-23 shows the use of an inside micrometer on a large hole.

Inside micrometer calipers are designed with caliper jaws that permit quick, direct inside measurements with micrometer accuracy, Figure 7-24.

Using The Inside Micrometer

One contact point is held or rested in a fixed position. The other contact is side wiggled as it is adjusted. This pivot action is necessary to be sure the tool is spanning the true diameter of the inside hole, Figure 7-25.

VERNIER CALIPERS

The Vernier caliper, Figure 7-26 consists of a graduated steel rule or bar. One end of the bar has a fixed jaw or contact point. A

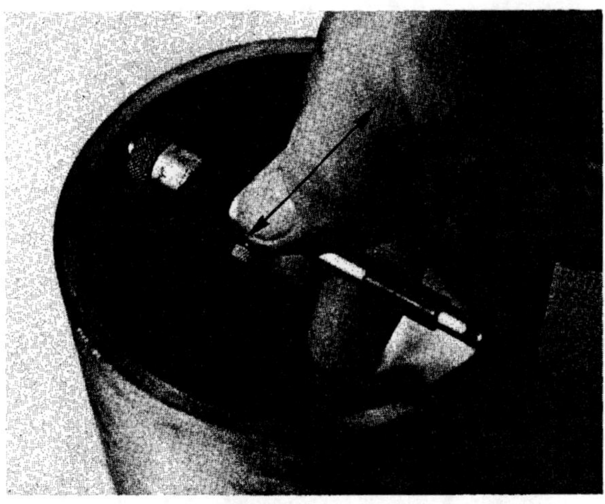

Fig. 7-25 Pivoting the inside micrometer (horizontally) to obtain the longest measurement

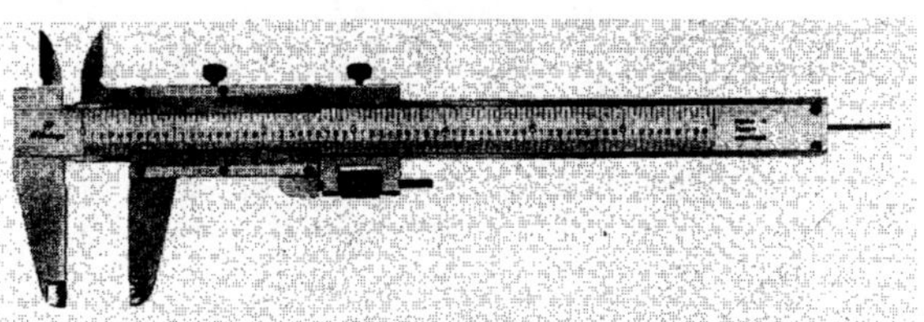

Fig. 7-26 Vernier caliper with inside, outside, and depth measurements (MTI Corporation)

movable jaw sliding on the bar carries a graduated Vernier plate. The plate is arranged so that it may be compared with the fixed scale on the bar. Figure 7-27 shows a 25-division Vernier. The bar of the tool is graduated in 40ths, or 0.025 of an inch. Every 4th division is numbered and represents 1/10 of an inch. Some calipers are a 50-division Vernier which makes the reading easier.

On the Vernier plate, there is a space divided into 25 parts and numbered from 0 to 25. The 25 divisions on the Vernier occupy the same space as 24 divisions on the bar. Therefore, the difference between one of the 25 spaces on the Vernier and one of the 24 spaces on the bar is 1/25 or 1/40 or 1/1000 of an inch.

Note: This difference in scale spaces is the basic measuring principle of a Vernier. The Vernier caliper is capable of reading to an accuracy of 0.001 inch.

When the 0 line on the Vernier matches with the 0 line on the bar, the reading is an even 0.025 inch. As each line past the 0 line of the Vernier coincides with the next line of the bar, the difference continues to increase. The increase is 1/1000 of an inch for each division until the line 25 on the Vernier coincides with the 24th line on the bar.

At this point, the 0 line of the Vernier has passed another 0.025-inch graduation on the bar.

Reading Vernier Calipers

Example: To read the Vernier caliper in Figure 7-27:

Count the inches shown . . . . .   1.000 inch
Count the tenth's showing
   between the inch and 0
   line on the Vernier . . . . . .   0.400 inch
Count the 0.025 divisions
   showing after the tenth's
   and before the 0 line of
   the Vernier . . . .   0.025 inch
Add the number of divisions
   on the Vernier from 0 to a
   line which exactly coincides
   with a line on the bar . . . . .   <u>0.011 inch</u>
The total reading is . . . .   1.436 inches

METRIC VERNIER CALIPERS

The same principles are used in reading metric Vernier tools as those in English measure, except that readings are obtained in 1/50 of a millimetre or 0.02 mm. The bar is graduated in centimetres, millimetres and half millimetres. The Vernier gives a final reading in

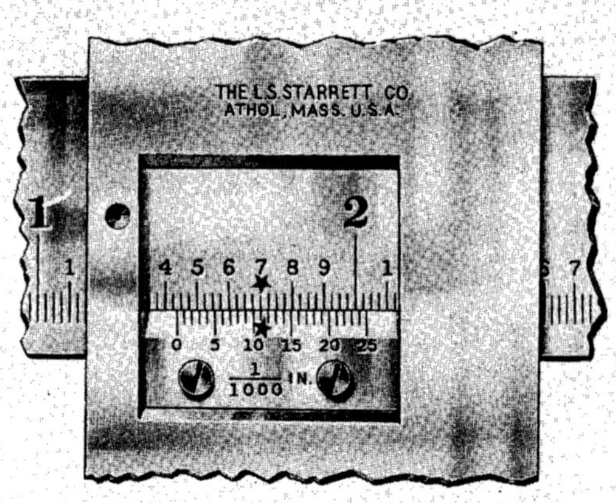

Fig. 7-27 A 25-division Vernier
(L. S. Starrett Company)

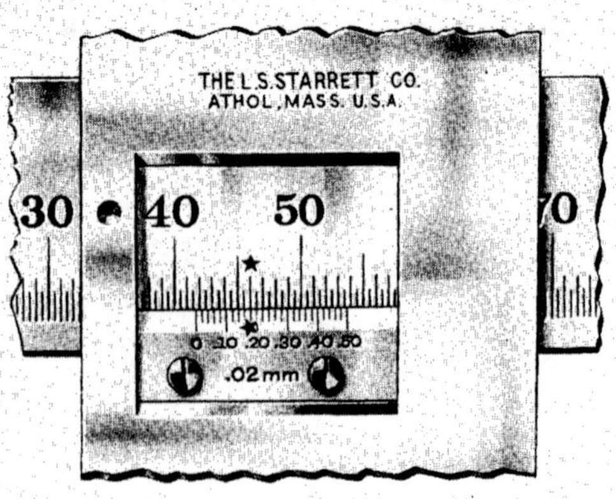

Fig. 7-28 Reading the metric Vernier caliper
(L. S. Starrett Company)

Fig. 7-29 Vernier caliper being used for an outside measurement (L. S. Starrett Company)

1/50 or 00.2 mm. Each bar graduation is 0.05 mm. Every 20th graduation is numbered in sequence (10 mm, 20 mm, 30 mm) over the full range of the bar. This provides for direct reading in millimetres. The Vernier plate is graduated in 25 parts, each line representing 0.02 mm. Every 5th line is numbered in sequence (0.10 mm, 0.20 mm, 0.30 mm) providing for direct reading in 0.02 mm.

Reading Metric Vernier Calipers

Example: Proceed as follows to read the metric Vernier caliper shown in Figure 7-28.

Count the millimetres that lie
  between the 0 line on the bar
  and the 0 line of the Vernier
  plate . . . .   41.5 mm
Find the graduation on the
  Vernier plate that coincides
  with a line on the bar and
  note its value in 0.02 mm.
  (Shown by the stars) . . . .   <u>0.18 mm</u>
Add this amount to the mil-
  limetres shown on the bar.
  The total reading is . . . .   41.68 mm

The Vernier caliper is equipped with a fine adjusting screw to obtain precision measurement. This caliper makes both inside and

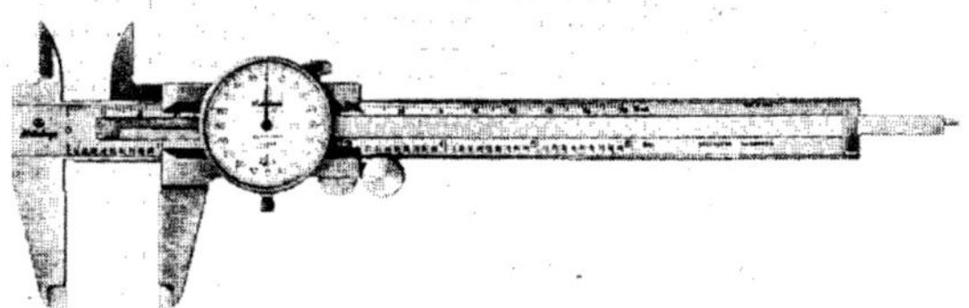

Fig. 7-30 Slide caliper with direct-dial reading (MTI Corporation)

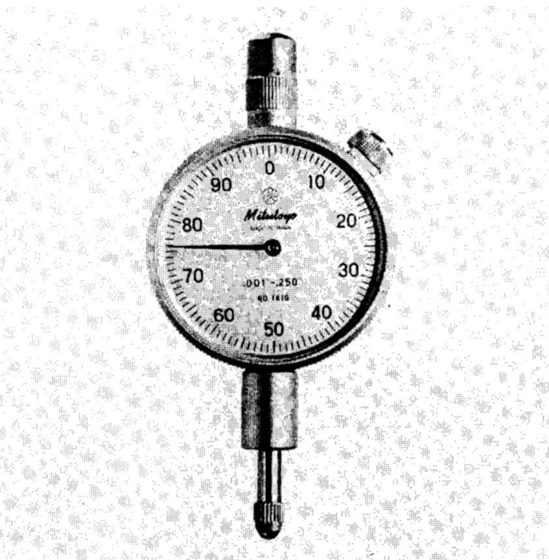

Fig. 7-31 Dial indicator (MTI Corporation)

outside measurements, Figure 7-29. Knurled clamp screws are provided to lock the adjusting bracket when making fine adjustment and lock the movable jaw once it is set to size. Care must be taken when measuring with the Vernier caliper. A good sense of feel must be used to obtain accurate measurement. After the caliper is locked, recheck the measurement to be sure no additional movement of the movable jaw was made when locking.

Slide calipers are now available with direct dial readings, Figure 7-30. These tools provide quick, simple reading with no calculations. Metric-measuring *dial calipers* are also available.

DIAL INDICATORS

The dial indicator, Figure 7-31, is one of the most widely used instruments today in

Fig. 7-32 Magnetic base for dial indicator (MTI Corporation)

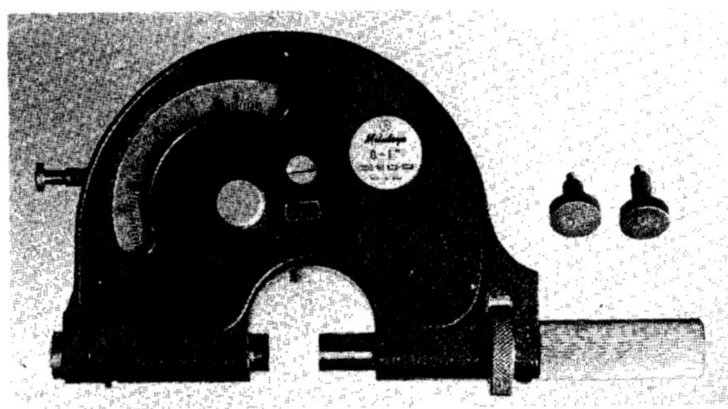

Fig. 7-33 Dial snap gage (MTI Corporation)

Fig. 7-34 Standard dial depth gage (Chicago Dial Indicator Company)

measurement, layout, inspection, and quality-control operations. Dial indicators are made with all the precision of a fine watch so that accurate measurements in either inch or metric readings are possible. The magnetic base provides a convenient holder for the indicator, Figure 7-32.

Graduations are available for reading in 0.001, 0.005, 0.0025, 0.0001, and 0.00005 inch, with travel ranges from 0.003 inch to 12 inches. Many types and variations of dial indicators are available. The regular contact points can also be replaced with contacts of almost any shape or length to suit the work. Rubber dust guards are furnished to seal out dust and foreign matter.

OTHER PRECISION GAGES

Dial snap gages, Figure 7-33, are of the *go* and the *no go* type. The gage will *go* or fit the correct dimension and will *no go* or not fit if the part is not the correct dimension. *Dial gages* are direct read from a pointer and graduated dial to provide accuracy and speed in many inspection operations of inch or metric parts, Figure 7-34.

The *dial comparator* is used in the same manner as the bench micrometer, Figure 7-35. It is used to inspect duplicated parts and various dimensions either in bench inspection or on the production line.

Fig. 7-35 Dial comparator (Chicago Dial Indicator Company)

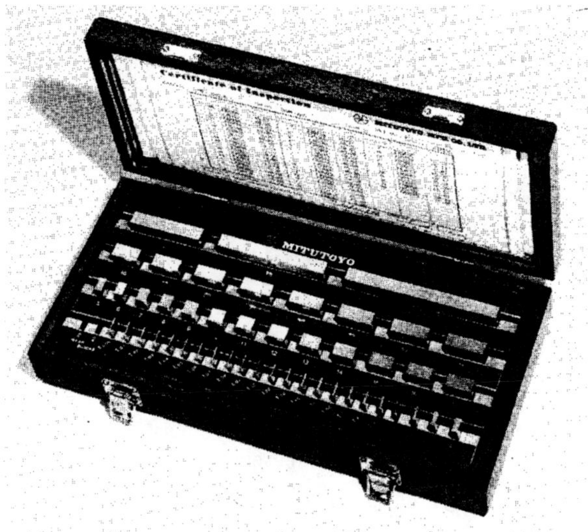

Fig. 7-36 Set of standard gage blocks
(MTI Corporation)

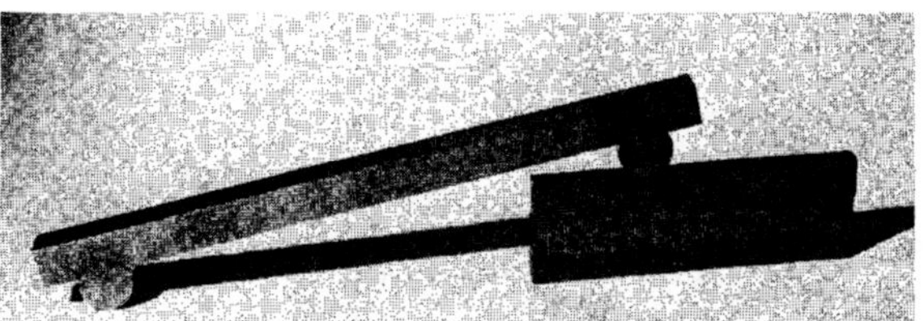

Fig. 7-37 Sine bar raised
to an angle on precision parallels

Fig. 7-38 Mechanics improved level
(L. S. Starrett Company)

Gage blocks are the industrial standards of length used for standardizing precision measuring instruments, Figure 7-36. They are precision rectangular blocks of hard and stable material with measuring or gaging surfaces on each end. To obtain a desired exact length, blocks of different lengths are selected and *wrung together* to form a stack of blocks. To maintain dependable accuracy, gage blocks must be kept free of dirt, moisture, and dust.

The *sine bar* is a precision bar made of stabilized, highly finished chrome steel, Figure 7-37. The bar rests on two cylindrical rolls, the centers of which are usually 5 inches or 10 inches apart. By the use of gage blocks, the sine bar can be built up or set to any desired angle in relation to a flat, true base surface.

LEVELS

Precision levels are made to set up and test the correct level positions of machinery, Figure 7-38.

Many more precision tools are made to aid the benchworker in the task of precision measuring and inspection. Precision measuring and inspection requires patience and exactness to the required degree of accuracy. Take time to measure and check precision work carefully. Tools must be properly used to obtain accurate measurement.

REVIEW QUESTIONS

A. Multiple Choice

1. Which of the following best describes the way to hold precision measuring tools to obtain accurate measurements?
    a. Loosely
    b. Rigidly
    c. Delicately firm
    d. Strongly

2. The number of threads per inch on the English micrometer spindle is:
    a. 25 threads per inch.
    b. 50 threads per inch.
    c. 40 threads per inch.
    d. 100 threads per inch.

3. The Vernier micrometer can be read to an accuracy of:
   a. 0.01 inch.   c. 0.0001 inch.
   b. 0.001 inch.   d. 0.00001 inch.

4. The Vernier caliper can be read to an accuracy of:
   a. 0.01 inch.   c. 0.0001 inch.
   b. 0.001 inch.   d. 0.00001 inch

B.   Short Answer

6. Read the following micrometer settings. Place your answers in the spaces below the drawings.

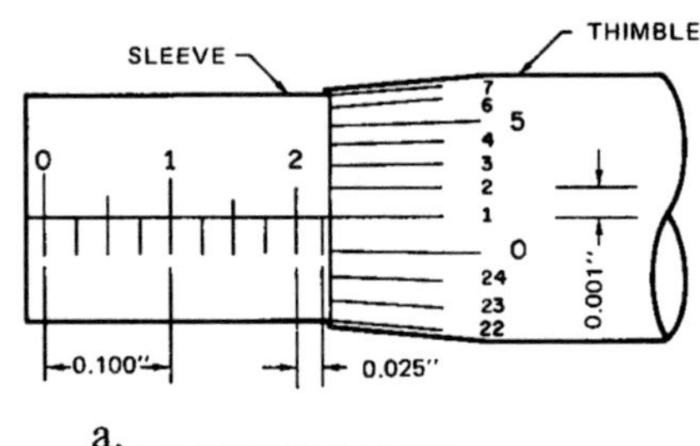

a. ____________________

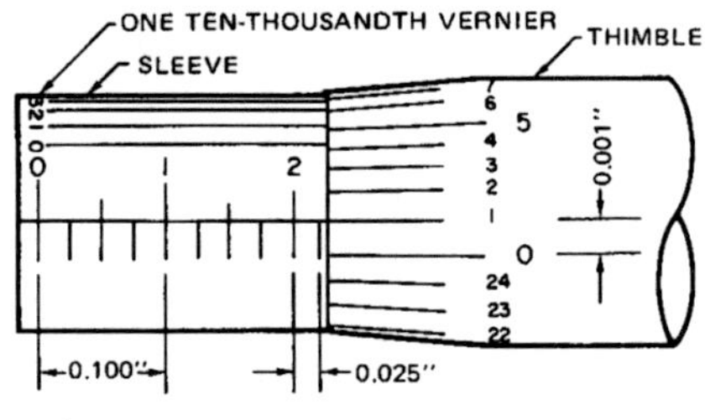

b. ____________________

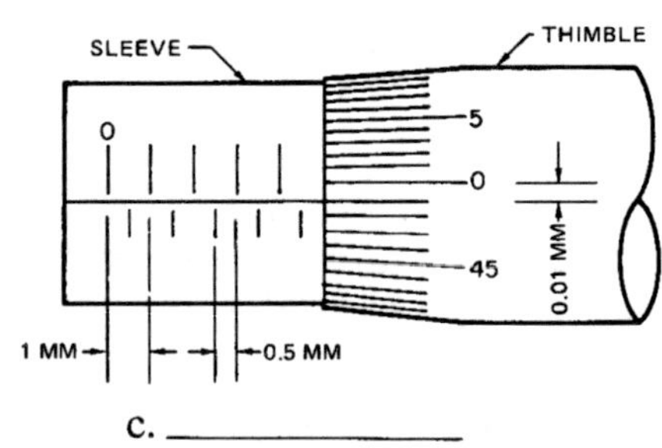

c. ____________________

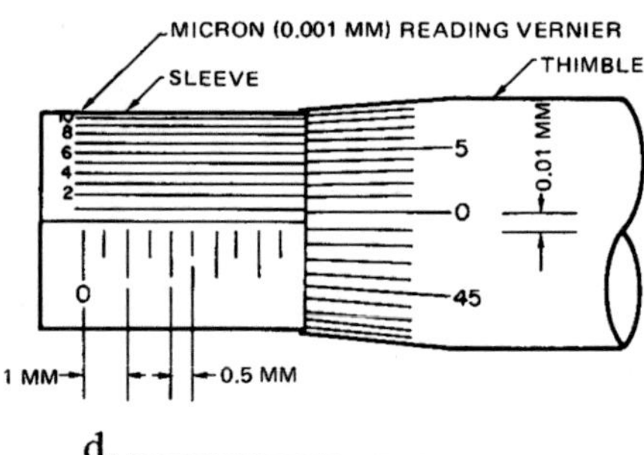

d. ____________________

7. Read the following Vernier caliper settings. Choose the correct answer from the choices listed below the drawings.

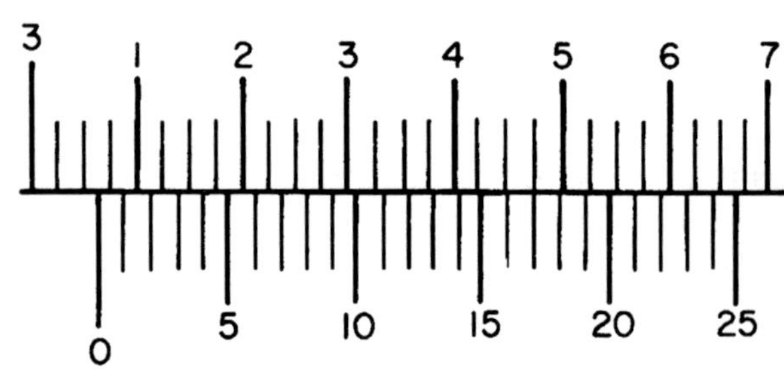

1. 3.475 INCHES
2. 3.067 INCHES
3. 3.050 INCHES

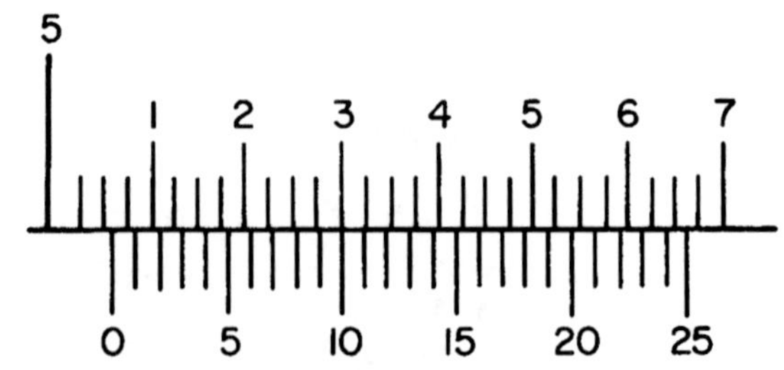

1. 5.300 INCHES
2. 5.050 INCHES
3. 5.060 INCHES

8. List the six main parts of a standard micrometer caliper.

9. Explain what is meant by the term *discrimination* in relation to measuring tools.

10. Another common shop name for a micrometer caliper is a _________.

11. What is the difference between a standard micrometer and a Vernier micrometer?

12. When measuring with the micrometer, what is the first procedure to follow before doing anything else?

13. The Vernier caliper is capable of reading to _________ inch.

ACTIVITIES

1. From at least four measuring tools provided by your instructor, state the following information for each tool:
   a. Tool name
   b. Measuring capability, direct or indirect
   c. Check the tool with a standard and reset if necessary
   d. Identify the discrimination of the tool

2. For each of the four measuring tools provided, accurately measure a part or parts to a given required accuracy. Select the proper tool to obtain the required accuracy, using both English and metric measurement.

UNIT 8 TOLERANCES AND FITS

OBJECTIVES

After completing this unit, the student will be able to

- define fitting as applied to metalwork.

- identify the three general groups of fits discussed.

- define the terms actual size, basic size, limits of size, and tolerance.

INTRODUCTION

In machine construction, a certain amount of hand fitting is required. This fitting makes the part surfaces contact as they should. Proper fitting of part surfaces is important so that the machine or mechanism operates according to design.

In metalworking, the term *fit* refers to the relation the surface of one part has to another. Parts may be required to contact each other in certain ways. The act of causing the required fit or relation of part surfaces is called *fitting*. This operation is done at the bench by the benchworker.

FITTING INSTRUCTIONS

Parts should only be made to fit the specified print requirements. If the parts fit more or less than required, productive time is wasted. The desired fitting of parts is indicated in notes and dimensions on shop working drawings. These notes and dimensions are usually located near the title block, Figure 8-1. These are simply the size limitations or an allowable error from the basic size.

FITTING TERMINOLOGY

Some of the more common terms used in the shop related to fitting are:

- *Actual size* — the measured size.

- *Basic size* — the size from which allowable error (allowance and tolerance) is dimensioned.

- *Limits of size* — are the maximum and minimum allowable size a part can be made and still be acceptable for interchangeability and proper fit.

Tolerance

Tolerance is the total permissible error on a part. It is the difference between the maximum and minimum limits of size. On the

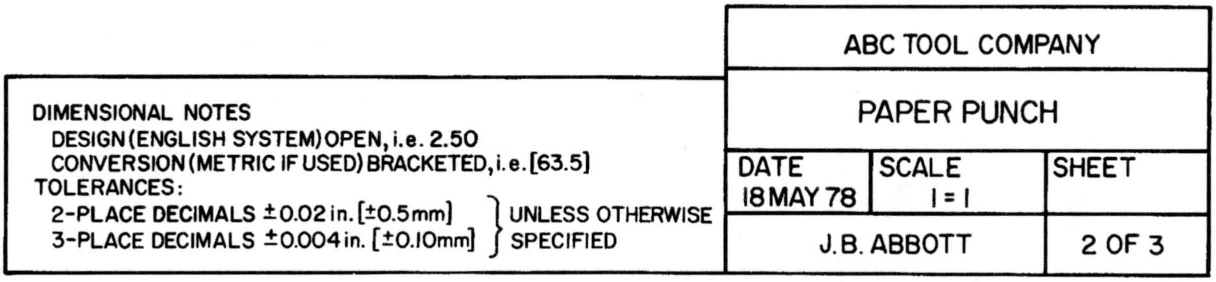

ABC TOOL COMPANY
PAPER PUNCH
DIMENSIONAL NOTES
DESIGN (ENGLISH SYSTEM) OPEN, i.e. 2.50
CONVERSION (METRIC IF USED) BRACKETED, i.e. [63.5]
TOLERANCES:
2-PLACE DECIMALS ±0.02 in. [±0.5mm]
3-PLACE DECIMALS ±0.004 in. [±0.10mm]
UNLESS OTHERWISE SPECIFIED
DATE
18 MAY 78
SCALE
1 = 1
SHEET
J.B. ABBOTT
2 OF 3

Fig. 8-1 Notes and dimensions taken from a working drawing

print, tolerance may be expressed by either unilateral tolerance or bilateral tolerance.

Unilateral tolerance means that the total error allowed is in one direction only. For example,

Basic Dimension	Tolerance
1.000″	- 0.002″
4.500″	+ 0.002″

These would be unilateral tolerances, since the total tolerance in each case is in one direction.

Example: A 1-inch basic size shaft with a minus 0.002-inch unilateral tolerance could vary in size from 0.998 inch to 1.000 inch. With the plus 0.002 inch in tolerance, the shaft size could vary from 4.500 inches to 4.502 inches, Figure 8-2.

Bilateral tolerance means that the total error allowed is divided partly plus, partly minus. For example,

$$1.000'' \quad \begin{matrix} +0.001'' \\ -0.001'' \end{matrix}$$

Example: The total tolerance of 0.001 inch is given in two directions, plus and minus. The 1-inch basic size shaft with the bilateral tolerance of ± 0.001 inch could vary in size from 0.999 inch to 1.001 inches, Figure 8-3.

Allowance

Allowance is an intentional difference between the maximum material size limits of mating parts. It is the maximum clearance or interference between mating parts. This

4.502
4.500
UNILATERAL TOLERANCE
4.500″ + 0.002″ -0.000″
1.000
0.998
UNILATERAL TOLERANCE
1.000″ +0.000″ -0.002″

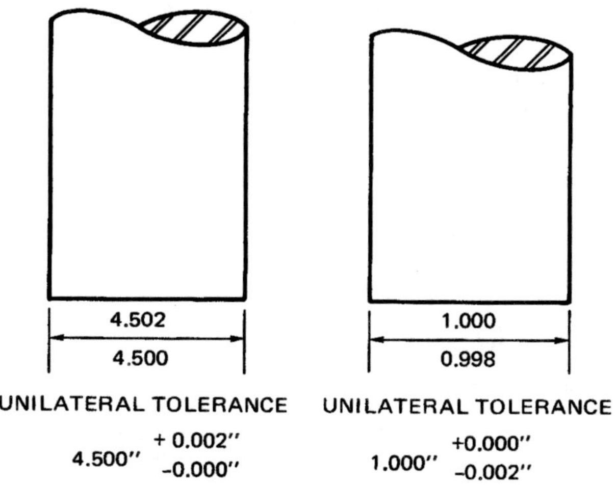

Fig. 8-2 Unilateral tolerance showing one-direction error

1.001
0.999
BILATERAL TOLERANCE
1.000″ ±0.001″

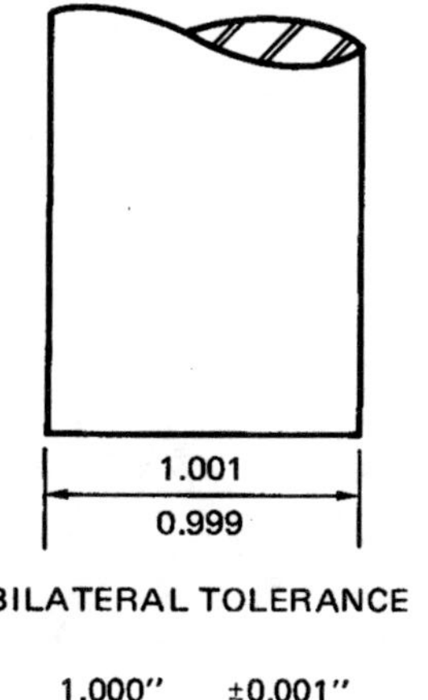

Fig. 8-3 Bilateral tolerance showing two-direction error from basic size

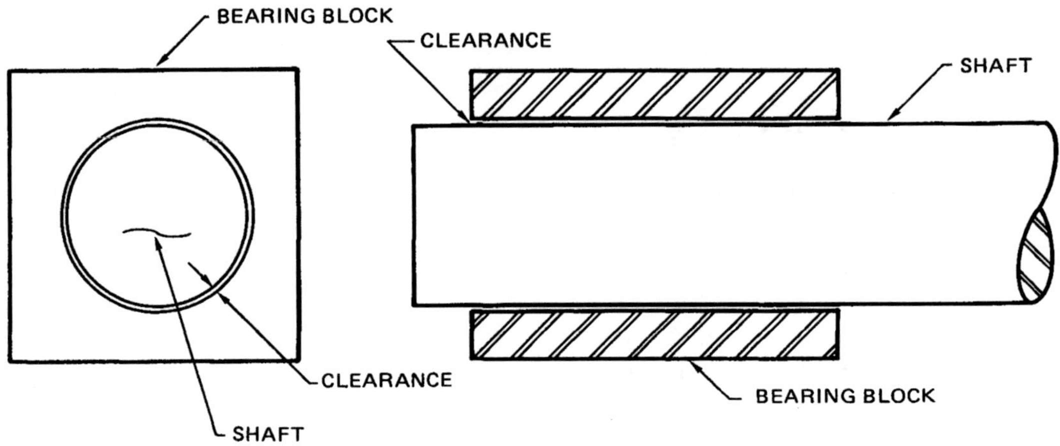

Fig. 8-4 Clearance fit showing tolerance between two parts

allowance provides for either a loose fit or a tight fit.

Fit

Fit is the range of tightness or looseness between two parts as a result of a specific combination of tolerances and allowances. The three types of fits are clearance, transition, and interference.

A *clearance fit* is a running fit which always allows clearance of a free fit between mating parts, Figure 8-4. This fit is used when one part turns inside another, such as a shaft turning inside a bearing. The fit must not be so tight that it keeps the shaft from turning.

The *sliding fit* is a clearance fit which has one part sliding on or against another part in a snug, close manner with some clearance, Figure 8-5. These parts are usually held together by means of adjustable contact strips. In some cases, the weight of one part on the other maintains the contact fit. The clearance and running fits probably require the most amount of hand-fitting at the bench to obtain the desired results. Alignment and contact are usually the limiting requirements.

A *transition fit* provides either a small amount of clearance or interference between the mating parts.

With an *interference fit*, the separate parts are forced or shrunk together so they act as one single part. This fit has limits of size which always provide for a tight fit. For ease of assembly, machining, or strength of part design, the parts are either forced or shrunk together. The three ways to do this are discussed below.

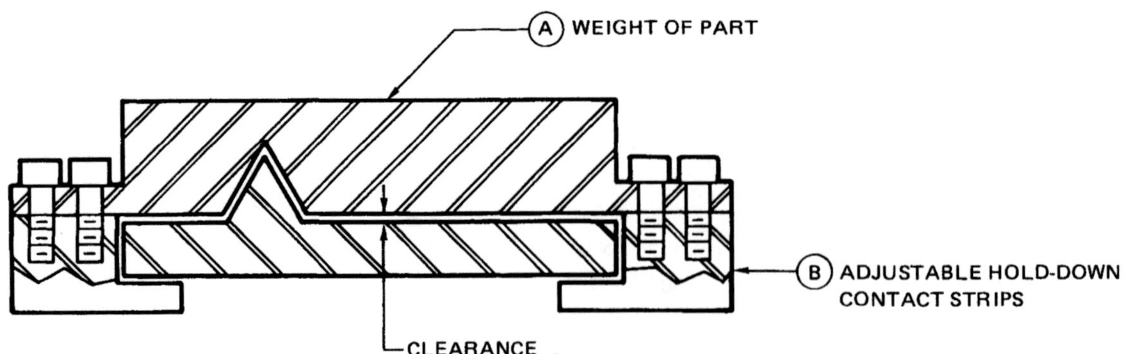

Fig. 8-5 Sliding fit: (A) Contact pressure maintained by weight of part,
and (B) contact pressure can also be maintained by adjustable contact strips or clamps

A *forced fit* is obtained by pressing one part onto another, Figure 8-6. Forced fit allowances are usually about 0.0015 inch per inch of diameter. The forced fit allowance is given to the shaft diameter, not to the hole where pins, shafts, etc., are to be forced into a hole, collar, or hub. The fit tolerance is given to the bushing hole where pins, shafts, etc., are run inside a sleeve bearing.

A *shrink fit* is obtained by heating the outside piece, placing it in the proper location to the inside part, and allowing it to shrink into position as it cools. A torch or furnace may be used to heat the part. A general rule for shrink fit allowance is 0.001 inch plus 0.001 inch for each inch of diameter of the part, Figure 8-7. This rule does not apply on thin-wall parts.

The *expansion fit* is used to locate one part permanently inside another, Figure 8-8. This method of fitting is used for ease of assembly. The inside part is shrunk by subzero cooling. It is then placed inside the outer part and allowed to expand into position as it picks up surrounding heat. The outside part may be slightly warmed to allow for a combination shrink and expansion fit.

Great care must be taken in the application and assembly of interference fits. Exact part dimensions from the print must be held to prevent part damage from breaking or bending.

ALLOWANCE MEASUREMENT SYSTEMS

The two systems of measurement used to identify the part receiving the allowance are the basic hole system and the basic shaft system.

The *basic hole system* is a system for fits in which the design of the hole is the basic size and the allowance is given to the shaft. The hole must be the basic size where standard tools are used to produce the hole, Figure 8-9. This provides for the greatest economy.

The *basic shaft system* is a system for fits in which the design of the shaft is the basic size and the allowance is applied to the hole,

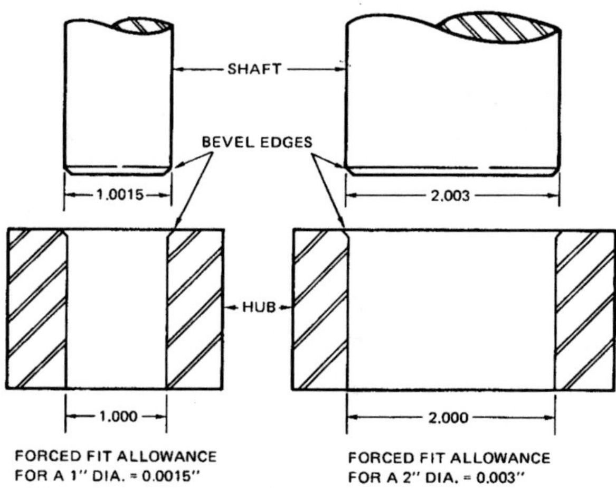

Fig. 8-6 Forced fits require interference between two mating parts — bevel edges for entrance ease

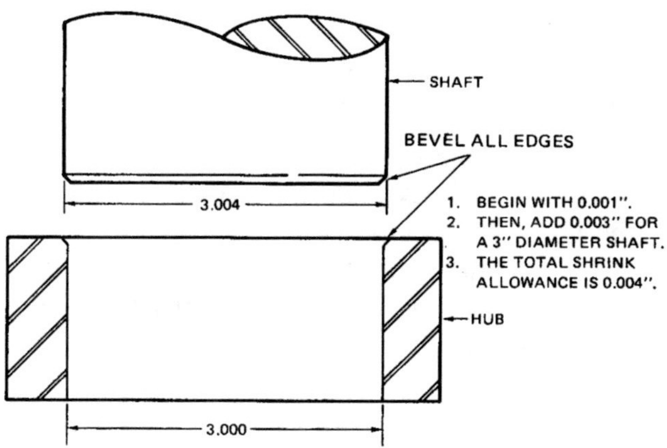

Fig. 8-7 Shrink fit showing calculation for shrink allowance of a 3-inch shaft fit

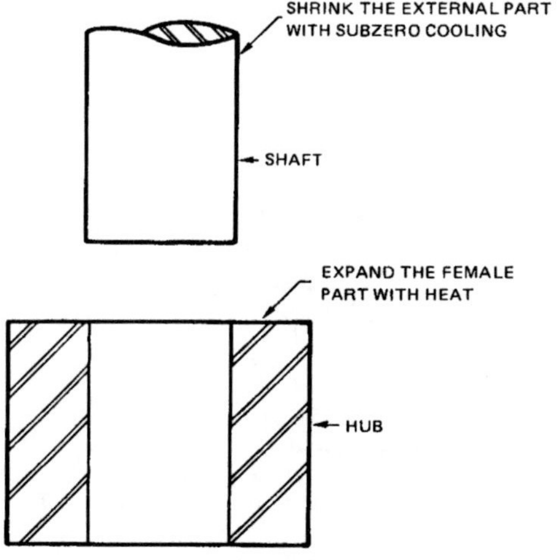

Fig. 8-8 Expansion fit

Figure 8-10. The shaft must be selected as basic size when using standard-size bars purchased from suppliers. This represents the greatest cost savings because the bar requires no further machining Special tools may be required to produce the hole in the mating part.

GENERALLY ACCEPTED TOLERANCES FOR FITTING

General tolerances apply whenever specific tolerances are not given on the drawing. When no tolerance is specified, the typical tolerance is ± 1/64 inch for fractional dimensions and ± 1/2 degree for angular dimensions. Decimal dimension tolerance is plus or minus one figure to the nearest significant figure. For example,

A 2-place decimal tolerance is ± 0.01 inch.
A 3-place decimal tolerance is ± 0.001 inch.

Allowable tolerances are usually given on the print. A good rule is to always work to the larger limit on outside-diameter work and to the smaller limit on inside-diameter work. It is easier to remove more material from the outside or inside of a part than it is to put back material if too much is removed.

PRINT SPECIFICATIONS

Look for the word accurate on the print where close plus or minus dimensions are to be held. This type of work may be on jigs and fixtures or on precision machine parts. Some industries and manufacturers set up their own accepted tolerance standards. The print title block or note section should state the fitting standards to be used.

Example:

2-place decimal — tolerance is ± 0.005 inch
3-place decimal — tolerance is ± 0.0015 inch
4-place decimal — tolerance is ± 0.0005 inch
5-place decimal — tolerance is ± 0.0002 inch

ALLOWANCES FOR FITS

When a shaft is fitted to a sleeve bearing, the inside diameter of the bearing is made a little larger than the shaft. This allows space for a thin film of oil between the surfaces, Figure 8-11. This clearance also allows for heat expansion of the shaft from the friction of the shaft turning in the bearing. The difference in diameters is called the allowance for the fit, or clearance.

Recommended allowances for different classes of fits are shown on the chart in Figure 8-12. The amount of allowance depends upon many of the following:

- Required part accuracy

- Size of the parts

- Kind of materials used in the parts

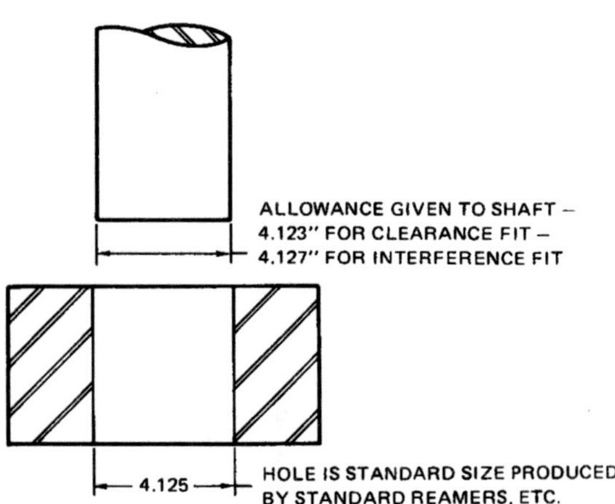

Fig. 8-9 Basic hole system of measure

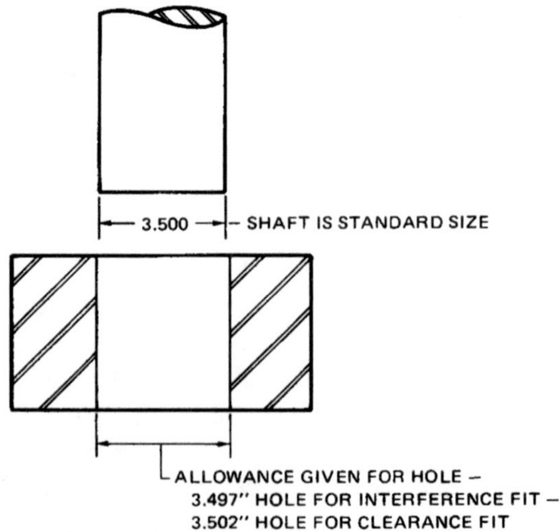

Fig. 8-10 Basic shaft system of measure

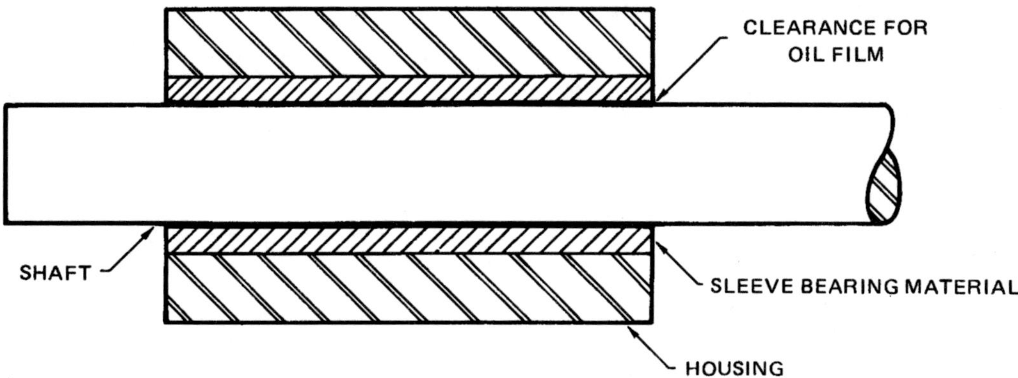

Fig. 8-11 Clearance shown for oil film between mating surfaces

Class		Up to ½ Inch	9⁄16–1 Inch	1 1⁄16–2 Inches	2 1⁄16–3 Inches	3 1⁄16–4 Inches	4 1⁄16–5 Inches
Tolerances in Standard Holes ★							
A	High Limit	+0.00025	+0.0005	+0.00075	+0.0010	+0.0010	+0.0010
	Low Limit	−0.00025	−0.00025	−0.00025	−0.0005	−0.0005	−0.0005
	Tolerance	0.0005	0.00075	0.0010	0.0015	0.0015	0.0015
B	High Limit	+0.0005	+0.00075	+0.0010	+0.00125	+0.0015	+0.00175
	Low Limit	−0.0005	−0.0005	−0.0005	−0.00075	−0.00075	−0.00075
	Tolerance	0.0010	0.00125	0.0015	0.0020	0.00225	0.0025
Allowances for Forced Fits							
F	High Limit	+0.0010	+0.0020	+0.0040	+0.0060	+0.0080	+0.0100
	Low Limit	+0.0005	+0.0015	+0.0030	+0.0045	+0.0060	+0.0080
	Tolerance	0.0005	0.0005	0.0010	0.0015	0.0020	0.0020
Allowances for Driving Fits							
D	High Limit	+0.0005	+0.0010	+0.0015	+0.0025	+0.0030	+0.0035
	Low Limit	+0.00025	+0.00075	+0.0010	+0.0015	+0.0020	+0.0025
	Tolerance	0.00025	0.00025	0.0005	0.0010	0.0010	0.0010
Allowances for Push Fits							
P	High Limit	−0.00025	−0.00025	−0.00025	−0.0005	−0.0005	−0.0005
	Low Limit	−0.00075	−0.00075	−0.00075	−0.0010	−0.0010	−0.0010
	Tolerance	0.0005	0.0005	0.0005	0.0005	0.0005	0.0005
Allowances for Running Fits ★★							
X	High Limit	−0.0010	−0.00125	−0.00175	−0.0020	−0.0025	−0.0030
	Low Limit	−0.0020	−0.00275	−0.0035	−0.00425	−0.0050	−0.00575
	Tolerance	0.0010	0.0015	0.00175	0.00225	0.0025	0.00275
Y	High Limit	−0.00075	−0.0010	−0.00125	−0.0015	−0.0020	−0.00225
	Low Limit	−0.00125	−0.0020	−0.0025	−0.0030	−0.0035	−0.0040
	Tolerance	0.0005	0.0010	0.00125	0.0015	0.0015	0.00175
Z	High Limit	−0.0005	−0.00075	−0.00075	−0.0010	−0.0010	−0.00125
	Low Limit	−0.00075	−0.00125	−0.0015	−0.0020	−0.00225	−0.0025
	Tolerance	0.00025	0.0005	0.00075	0.0010	0.00125	0.00125

Formulas for Determining Allowances

Class	High Limit	Low Limit	Class	High Limit	Low Limit
A	$+\sqrt{D} \times 0.0006$	$-\sqrt{D} \times 0.0003$	X	$-\sqrt{D} \times 0.00125$	$-\sqrt{D} \times 0.0025$
B	$+\sqrt{D} \times 0.0008$	$-\sqrt{D} \times 0.0004$	Y	$-\sqrt{D} \times 0.001$	$-\sqrt{D} \times 0.0018$
P	$-\sqrt{D} \times 0.0002$	$-\sqrt{D} \times 0.0006$	Z	$-\sqrt{D} \times 0.0005$	$-\sqrt{D} \times 0.001$

★Tolerance is provided for holes, which ordinary standard reamers can produce, in two grades, Classes A and B, the selection of which is a question for the user's decision and dependent upon the quality of the work required; some prefer to use Class A as working limits and Class B as inspection limits.

★★Running fits, which are the most commonly required, are divided into three grades: Class X, for engine and other work where easy fits are wanted; Class Y, for high speeds and good average machine work; Class Z, for fine tool work.

Fig. 8-12 Chart of allowances for different classes of fits (L. S. Starrett Company)

- Shape and thickness of the part materials
- Surface finish of the shaft and the hole
- Length and diameter of the fitting parts
- Type of lubricant used
- Speed of the mating parts
- Room temperature where the parts will be operating

Information regarding the limits of tolerance for the parts should be clearly shown on the drawing. If not, they should definitely be covered by written specifications to which reference must be made by notations on the drawing.

Thickness or feeler gages, Figure 8-13, are used to determine clearances between mating parts. They are made up of a number of thin blades, each marked with its thickness. The blade is inserted between the mating parts, and the thickness is read on the blade, Figure 8-14.

FITTING OPERATIONS

Fitting at the bench may need to be done in one of the following ways:

- Sawing — cutting material with hand or power band saw.
- Chipping — removal of material with a hammer and chisel.
- Filing — removal of material with files.
- Scraping — removal of small amounts of material with scraping tools.

  This is usually the final fitting process.
- Polishing — removal of material with fine abrasive materials.
- Grinding — removal of material with a grinding wheel.
- Cutting by machine — removal of material with machine tools such as the lathe, milling machine, and drill press.
- Shimming — changing the position of one part to another by placing thin material between the parts. Shims are used to raise a part higher or project it out further from another part.

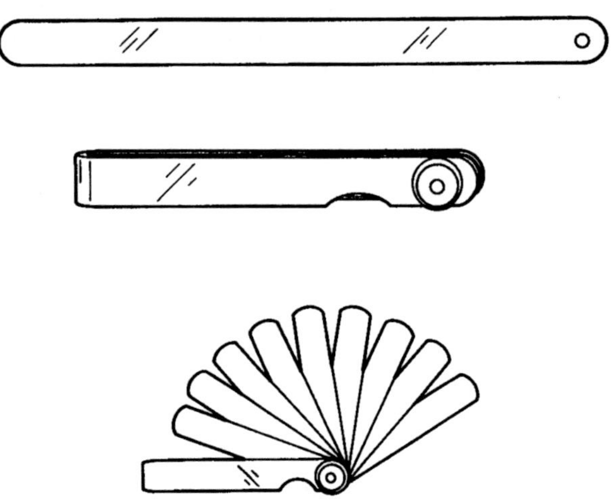

Fig. 8-13 Various feeler gages

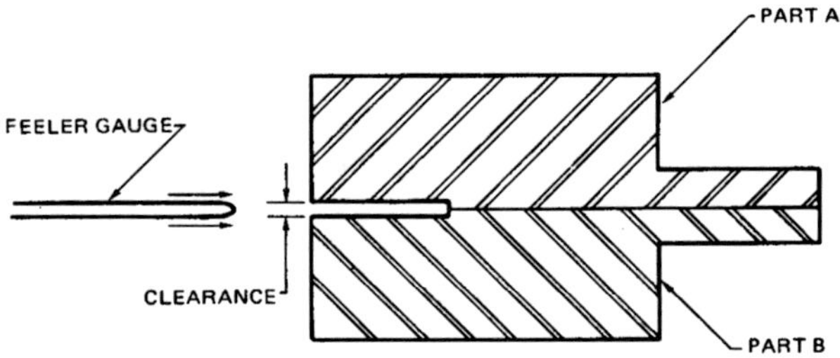

Fig. 8-14 Using feeler gage to measure or check clearance between fitted parts (A and B)

Good judgment, skill, and care are needed in proper print reading and fitting of parts. The life of the machine is determined by the fitting process. This process is literally in the hands of the benchworker as the proper fits are made. Fitting is best learned by working with an experienced craftsperson. Repair parts should be fit by consulting the repair manual as a guide to correct fitting.

Safety For Fitting Operations

- Read the print carefully to avoid costly mistakes.
- Do not operate power tools until the safe and proper operation of them is known.
- When sawing, be careful not to get cut on the sharp teeth of the saw or the ragged edges left by the sawing operation.

Fig. 8-15 Use approved eye protection when chipping or grinding (The Stanley Works)

- Be careful of sharp tools and sharp burrs left by cutting.
- Make sure fingers do not get pinched between parts.
- Protect finished surfaces from rust and nicks.
- Wear proper eye and face protection when chipping and grinding, Figure 8-15.
- Always use the file with a properly fitted handle.
- Replace tools and materials to their proper storage places immediately after use.
- Maintain a clean, well-organized work area.

REVIEW QUESTIONS

A. Multiple Choice

1. Fitting parts to closer tolerances than required is:
   a. seldom done.
   b. always done.
   c. wasteful.
   d. never done.

2. The relation one part surface has to another part surface is called:
   a. size.
   b. actual size.
   c. fit.
   d. scrape.

3. The operation of causing a certain relation of one part to another is called:
   a. sizing.
   b. fitting .
   c. scraping.
   d. surfacing.

4. Parts must be machined and fitted to size limitations to provide which of the following?
   a. Interchangeability
   b. Allowance
   c. Tolerance
   d. Exchange

5. Sliding fits are sometimes held together by which of the following?
   a. Interference
   b. Clearance
   c. Transition
   d. Weight of one part on the other

6. Which of the following fits generally requires the most amount of hand fitting?
   a. Forced fits
   b. Clearance fits
   c. Expansion fits
   d. Shrink fits

7. When pins or shafts are to be force fitted into a hole, the force fit allowance is normally given as the:
   a. hole diameter.
   b. shaft diameter.
   c. shaft length
   d. surface finish.

B.   Short Answer

8. Name the two systems of measurement used to identify which part is to receive the allowance.

9. Define the term *tolerance*.

ACTIVITIES

1. From a selection of various fits provided by your instructor, identify the following types of fits:
    a. Clearance (running, sliding)
    b. Transition (clearance, interference)
    c. Interference (forced, shrunk, expansion)

2. From a shop print provided by your instructor, identify the various tolerances and allowances. Also locate the special notes that refer to fitting-accuracy instructions.

SECTION III
LAYOUT PRINCIPLES
UNIT 9 NONPRECISION AND SEMIPRECISION LAYOUT TOOLS

OBJECTIVES

After completing this unit, the student will be able to

- identify the common nonprecision and semiprecision layout tools used in the shop.

- demonstrate the proper and safe use of nonprecision and semiprecision layout tools.

INTRODUCTION

The tools used for making a layout or drawing on a workpiece are called *layout* tools. Many of these tools must scratch or cut lightly into the workpiece to show visible marks. The accuracy of the job depends upon the proper use and care of these tools. The tools discussed in this unit are most commonly used in non-precision and semiprecision layout.

SAFETY PRECAUTIONS AND TOOL CARE

- Many of the tools used in layout have very sharp points. Keep the points sharp and protected.

- Care must also be taken to avoid injury caused by improper handling of these sharp, pointed tools. Do not carry sharp, pointed tools in your pockets.

- Never use a prick punch or center punch with a mushroomed head. Hard, brittle pieces of the punch can fly off and cause personal injury.

- Keep your tools adjusted and lubricated.

- Store them in a clean, dry area.

NONPRECISION LAYOUT TOOLS

Scriber

The scriber is probably the most frequently used of the marking layout tools. Scribers are made of tool steel with hardened points.

Two common styles are the straight single-end scriber and the double-end scriber, Figure 9-1A and B. With the double-end scriber, one end is bent 90 degrees to scribe in holes or other hard to reach areas. The straight scriber often has a magnet built into the opposite end of the scriber point. The scriber point is removable and stores in the handle for safe carrying. Scribers are made from 6 inches to 10 inches in length and approximately 3/16 inch in diameter.

How To Use The Scriber

The tool is held like a pencil to scribe or draw lines on the work surface. The handle of the scriber leans toward the direction in which it is moved. The scriber leans away from the straightedge so the tip marks as close as possible to the rule, Figure 9-2. This procedure produces a smooth, accurate line. Never use the scribe as a punch.

Layout Hammer

The *layout hammer* is used to strike punches, Figure 9-3. This produces layout marks required for general layout. The built-in magnifying lens aids the worker in aligning the point of the punch with fine cross lines of the layout. A standard ball peen hammer may also be used as a striking tool for punches, Figure 9-4.

Prick Punch and Center Punch

The *prick punch*, Figure 9-5, is a layout tool made of tool steel, about 4 inches to 6 inches long. Both ends of the prick punch are heat-treated. The point end is ground to an included angle of 30 degrees to 60 degrees. This tool is used to make the first small punch marks on layout lines or at center points where lines cross one another. This mark provides a place for the point of a divider to rest while scribing circles or arcs. Punch marks are sometimes called *witness marks*. This is because the shallow marks remain as a guide if the layout lines are rubbed off the work surface. Prick punch marks are also used to mark centers for small drilled holes.

The *center punch* is similar to the prick punch, except the point is ground to between 60 and 90 degrees, Figure 9-6. This punch is used to enlarge the prick punch marks at line intersections where larger drilled holes are required. Some center-punch styles are made automatic with the striking part enclosed in the handle, Figure 9-7. A downward pressure on the handle releases the spring-loaded striking part and makes the punch mark. This punch does not require the use of a hammer.

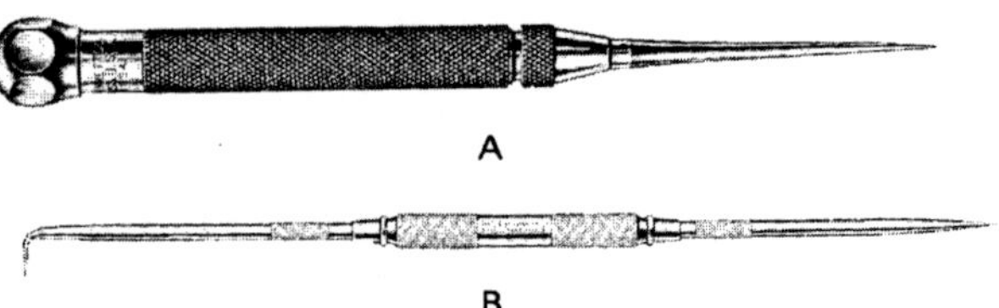

Fig. 9-1 (A) Straight, single-end scriber, and (B) double-end scriber (L. S. Starrett Company)

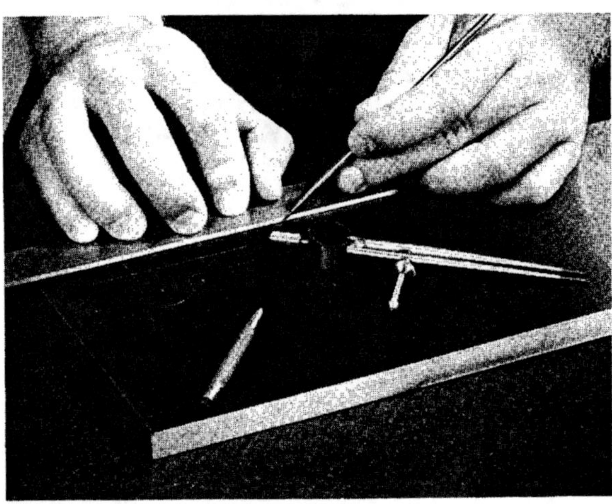

Fig. 9-2 Proper use of the scriber and the straightedge rule (DoAll Company)

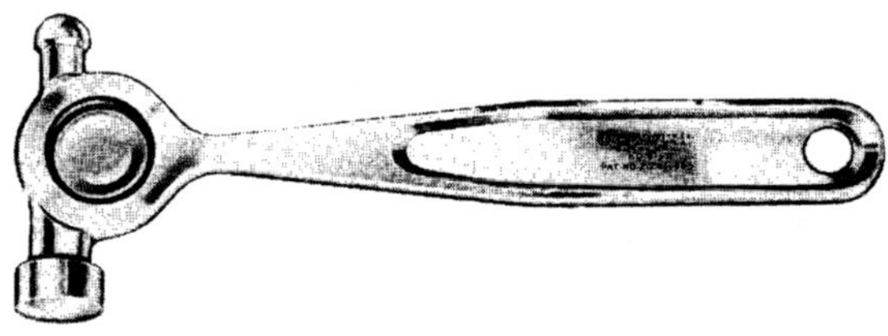

Fig. 9-3 Layout hammer with magnifying glass (L.S. Starrett Company)

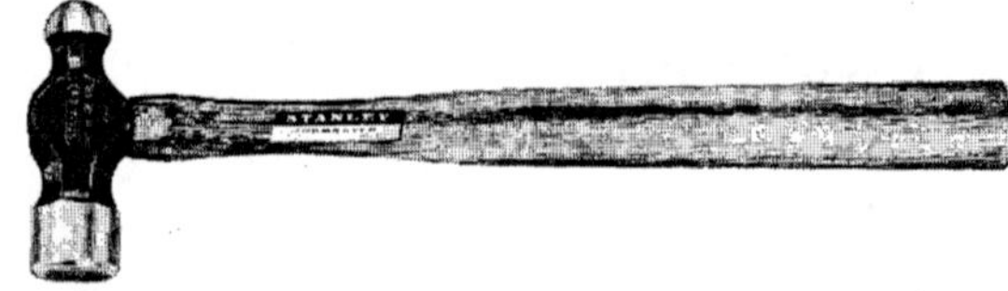

Fig. 9-4 Ball peen hammer used in layout operations (The Stanley Works)

Fig. 9-5 Prick punch (L. S. Starrett Company)

Fig. 9-6 Center punch (L. S. Starrett Company)

**Fig. 9-7 Automatic center punch
(L. S. Starrett Company)**

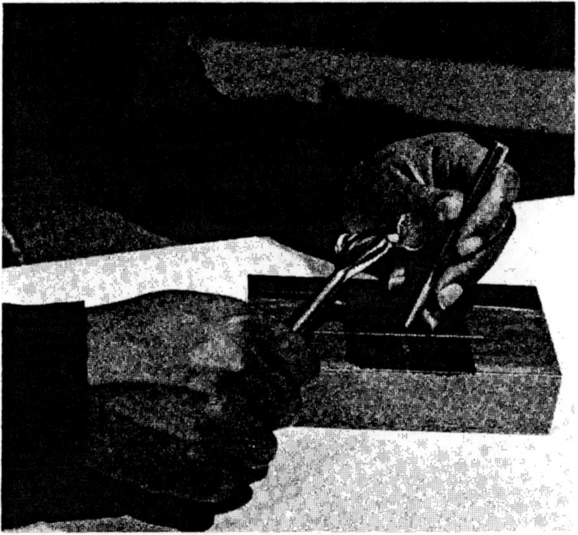

**Fig. 9-8 Tip the punch at an angle and set the point
on the intersection of the centerlines.**

How to Use Prick and Center Punches

1. Punches should always be kept sharp by dressing on a suitable grinder.
2. To set the punch on the intersecting lines, hold the punch as shown in Figure 9-8. Tip the punch to an angle and set the point at the intersection.
3. Bring the punch to a vertical position and tap gently with a hammer, Figure 9-9A.
4. Check the punch mark position for accuracy.
5. Correct, if necessary, by tipping the punch and striking again to move the punch mark to required location, Figure 9-9B.
6. Make additional layouts from this prick punch mark.
7. Enlarge the prick punch mark with a center punch used in the same manner.
8. Strike the center punch with more force to provide a good starting point for the drill bit.

C Clamps and Parallel Clamps

C Clamps are clamps made in the shape of a *C*, Figure 9-10. They are used to hold work to an angle plate, work table, or for clamping two pieces of work together. *Par-*

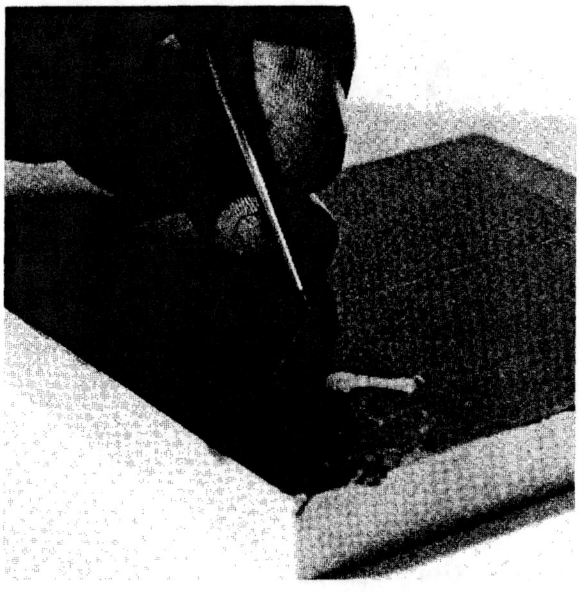

Fig. 9-9A Bring punch vertical and tap with hammer.

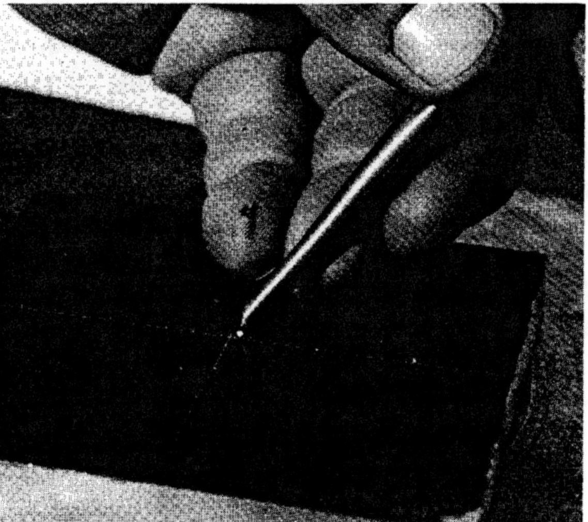

**Fig. 9-9B Tip punch and strike to move punch mark
to the cross of the centerlines.**

allel clamps are clamps having two jaws held together with adjusting screws, Figure 9-11. These clamps are extremely useful for holding work together in layout.

SEMIPRECISION LAYOUT TOOLS

These tools are considered accurate to within ± 1/64 inch (± 0.4 mm).

Transfer Punches

Transfer punches are used to transfer the center of existing holes in plates to other plates, Figure 9-12. They are made of tool steel, heat-treated, and ground to various standard diameters with a 90-degree, center punch point on one end.

How to Use Transfer Punches

1. Select the proper size diameter punch.
2. Place the punch through the existing hole.
3. Strike the head with a hammer.
4. The center of the existing hole is thus transferred to the new part.

Dividers

Dividers, Figure 9-13, are an indirect-reading layout tool used as a compass for scribing circles or arcs. Other uses include laying off distances of measurements taken from a steel rule and for measuring dimensions between lines or points. Dividers are adjustable and sized by the maximum opening between the points. The contacts are sharp, hardened points at the ends of straight legs. Close measurements are made by visual comparison rather than by feel. Dividers are limited in range by the opening of the legs. They are less accurate when the legs are fully open.

How to Set Dividers

1. Place one point in the 1-inch graduation line of the steel rule.
2. Adjust the other point until it splits the graduation line the required distance on the rule.

3. Distances are laid off by rotating the divider about its points, Figure 9-14.

Graduation lines on rules are made in V shape. This allows dividers to be set best by feel as the adjusting point centers in the V-shaped graduation of the rule. A magnifying glass may be used for more accurate settings of dividers.

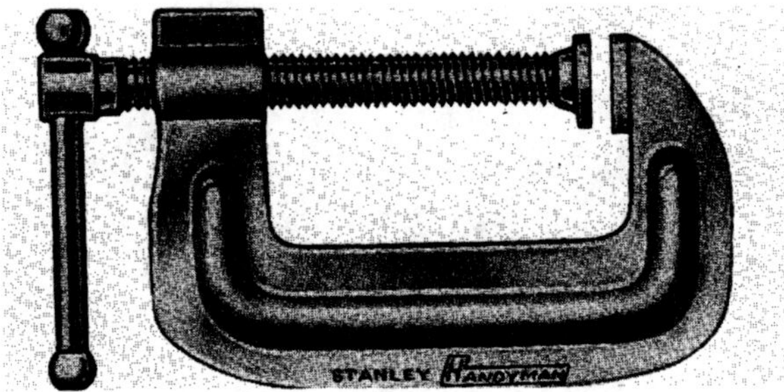

Fig. 9-10 C clamp (The Stanley Works)

Fig. 9-11 Parallel clamps

Fig. 9-12 Transfer punches used to transfer the center of a hole to another part

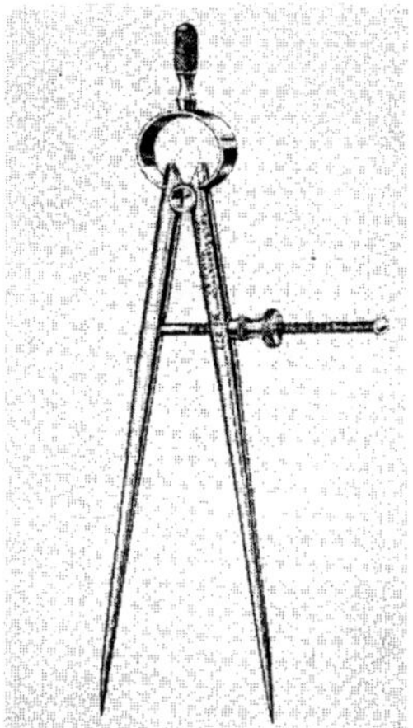

Fig. 9-13 Dividers are used to scribe circles or lay off distances. (L. S. Starrett Company)

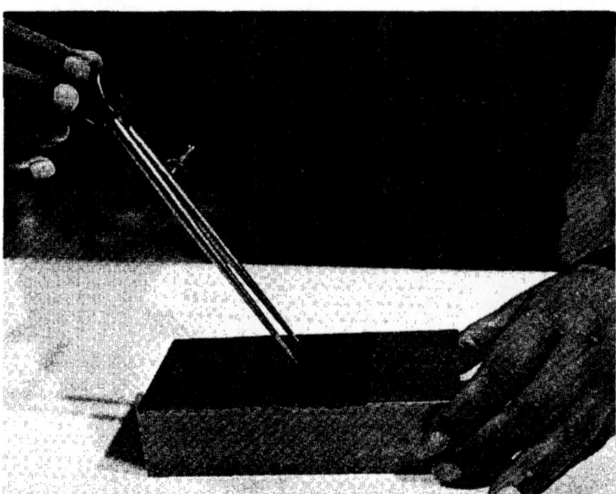

Fig. 9-14 Dividers used to lay off distances on a line

**Fig. 9-15 Dividers used to scribe
a circle on the workpiece**

How to Scribe A Circle Using Dividers

To scribe an arc or circle with dividers, Figure 9-15, use the following procedure:

1. Make a prick punch mark and insert one point of the dividers in the mark.
2. Grasp the knurled top of the tool with the thumb and forefinger and swing the scribing leg in an arc with downward pressure to cause a scribed mark on the material.
3. The scribing leg should be inclined in the direction of rotation to avoid the pivot point from slipping out of the prick punch mark. Both legs should be kept the same length.

Sharpening Divider Points

The points of the dividers, the scribe, and other pointed layout tools should be sharpened on an oilstone, Figure 9-16. The scriber points should be rotated between thumb and forefinger while stroking them back and forth on the oilstone.

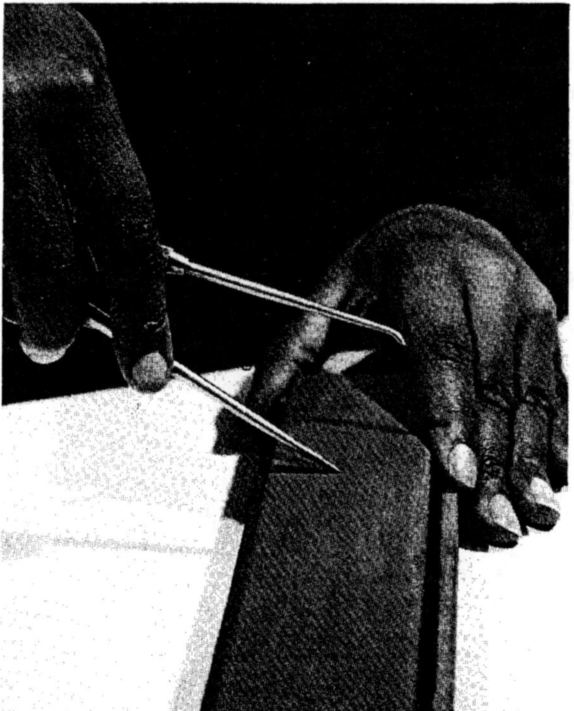

Fig. 9-16 The point of the calipers being sharpened to a point on an oilstone

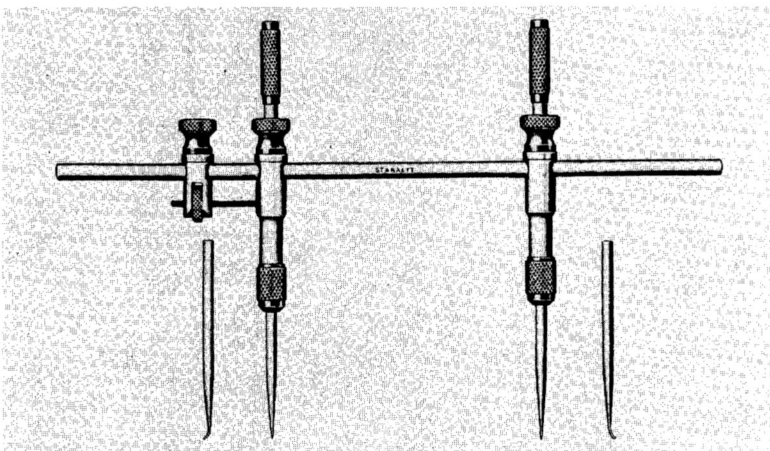

Fig. 9-17 Beam trammel (L. S. Starrett Company)

Beam Trammel

The *beam trammel* is used to scribe arcs and circles and do layout work beyond the range of the dividers, Figure 9-17. This tool is set to dimensions similar to the dividers except the points are adjusted and set along a bar or beam. Replaceable bent legs can be used for measuring with the trammel rather than the scribe points, Figure 9-18.

Hermaphrodite Caliper

The *hermaphrodite caliper*, or *morphy*, is a layout caliper having one bent guide leg and one straight scribing leg, Figure 9-19. The bent leg and the point leg should be adjusted to the same length. This tool is used to scribe lines parallel from an internal or external edge by reversing the bent leg as shown in Figure 9-20A and 9-20B. The removable leg of the caliper is shown in Figure 9-21.

How to Set Hermaphrodite Caliper

1. Hook the bent leg at the end of an accurate rule and adjust the pointed leg to the desired graduation, Figure 9-22.
2. With the leg in the reverse position, the tool is set by placing the end of the caliper even with the end of an accurate rule. Both the leg of the caliper and the rule end are placed against a straight surface, Figure 9-23.

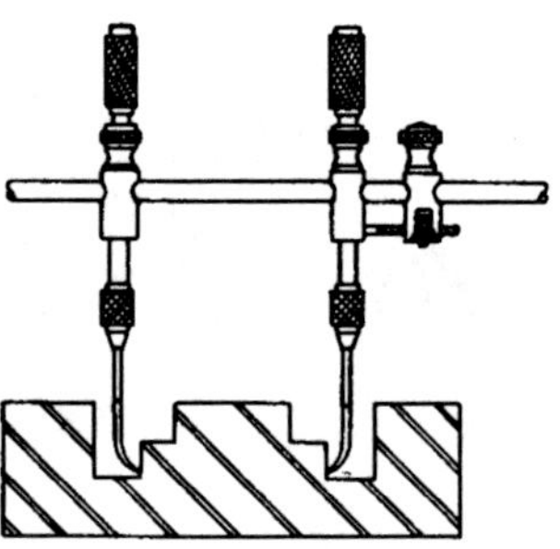

Fig. 9-18 Use of bent legs with the trammel to measure as an outside caliper (L. S. Starrett Company)

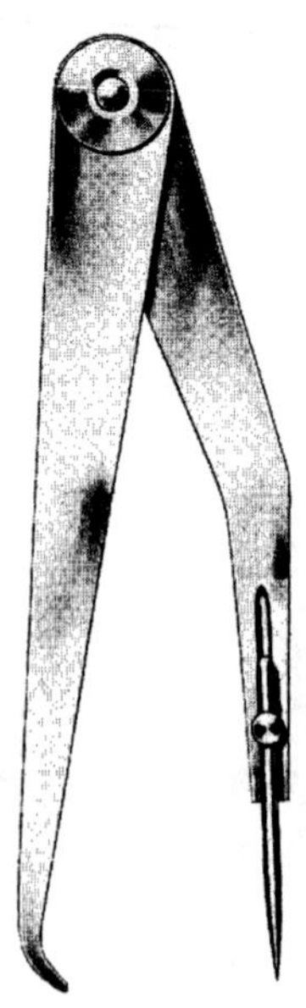

Fig. 9-19 Hermaphrodite caliper
(L. S. Starrett Company)

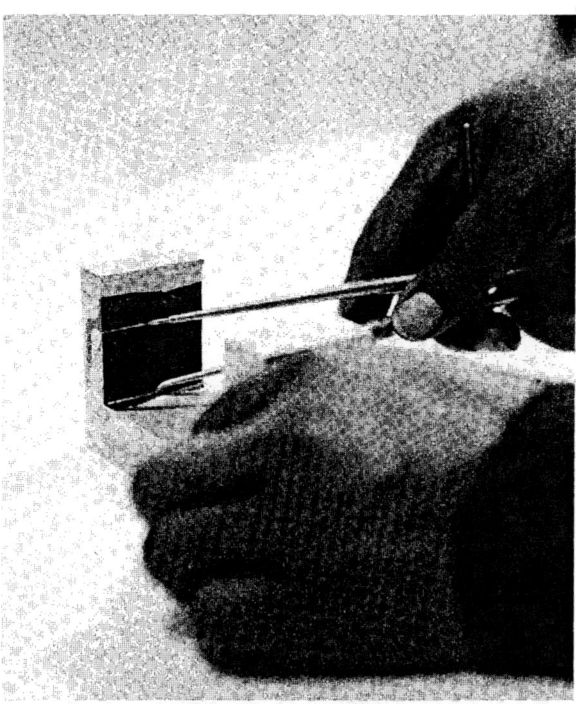

Fig. 9-20A Morphy used with bent leg out to scribe internal parallel lines

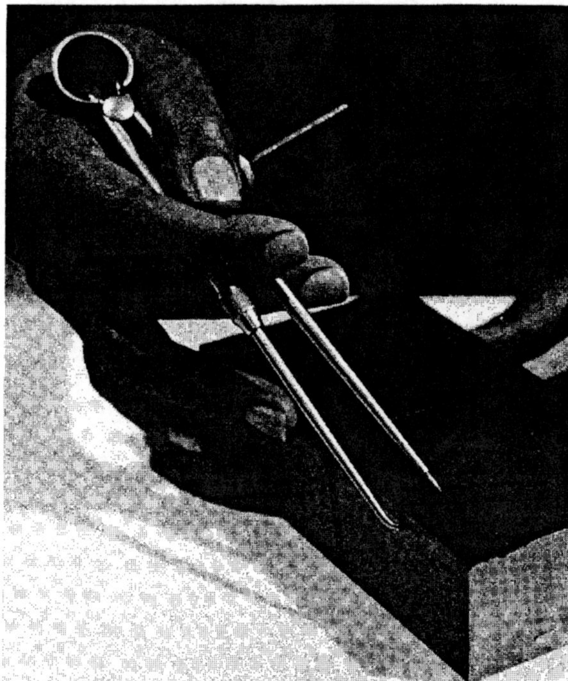

Fig. 9-20B Morphy used with bent leg in to scribe internal parallel lines

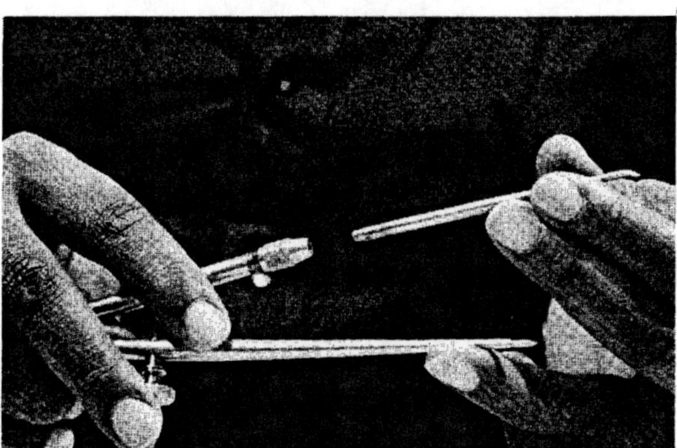

Fig. 9-21 Morphy reversible leg removed

3. The scriber leg is then adjusted to the desired measurement and the clamping nut tightened.

How to Use Hermaphrodite Caliper

1. Scribe the lines by exerting a slight pressure on the scriber and drawing the bent leg of the caliper along the surface or edge being scribed.
2. Lean the caliper in the direction in which the line is being scribed.

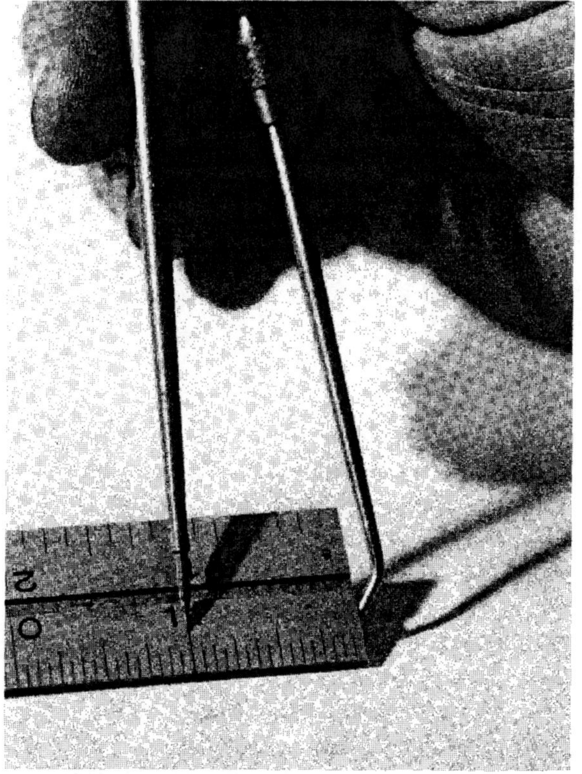

Fig. 9-22 Set the morphy caliper to a true rule for external layout.

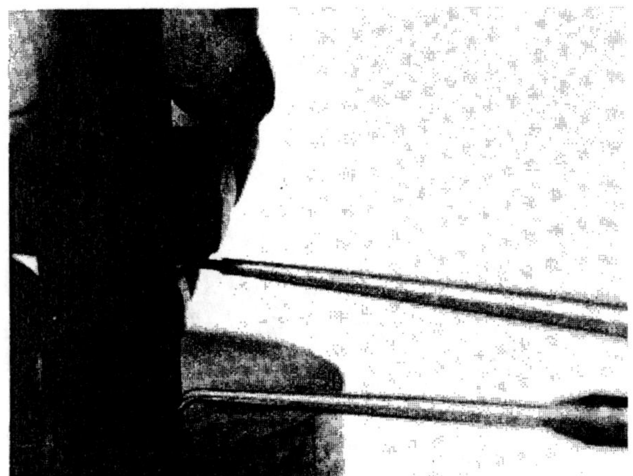

Fig. 9-23 Set the morphy caliper to a true rule for internal layout.

How to Locate Centers On Round Work

1. Set the caliper a little less than the radius of the work.
2. Hold the bent leg against and just below the work edge and scribe an arc, Figure 9-24.
3. Move the bent leg around the part at each 1/4 turn of the part and scribe the 4 arcs.
4. Center punch the center of the 4 arcs and check for accuracy.
5. The center punch mark can be moved by tipping the punch in the direction of the high side and repunching.
6. Retest the work and repunch until the desired accuracy is obtained.

Combination Set

The *combination set* is one of the most useful and versatile layout and measuring tools used in the shop. The set consists of four major parts, Figure 9-25. Some of its uses include checking inside and outside squareness, checking 45-degree angles, locating the center of a round workpiece, and drawing a line parallel to an edge.

The *steel rule* or blade may be fitted into the square head, center head, or the bevel protractor. At times, the blade may be used separately as a straightedge or for general mea-

Fig. 9-24 Location of the center on round work with the use of the morphy caliper

suring. Combination set rules are graduated in 8ths and 16ths on one side and in 32nds and 64ths on the other side. Rules are also graduated in various decimal parts of an inch and metric divisions.

The *square head* is used as a base guide to lay out lines parallel and at right angles to an edge. The blade is adjustable through the head and may be locked at a desired measurement to serve as a depth gage. The *spirit level* makes it possible to use as a simple level. The head contains a scribe for convenient use. Angles of 45 degrees and 90 degrees may be checked or laid out with the use of the head on the blade.

The *center head* is used to locate the center of round work. Square and hexagon work centers may also be found by using the center head.

The *bevel protractor* is used to lay out and check any angle from 0 to 180 degrees, Figure 9-26. The steel rule slides through the protractor and can be locked at any desired

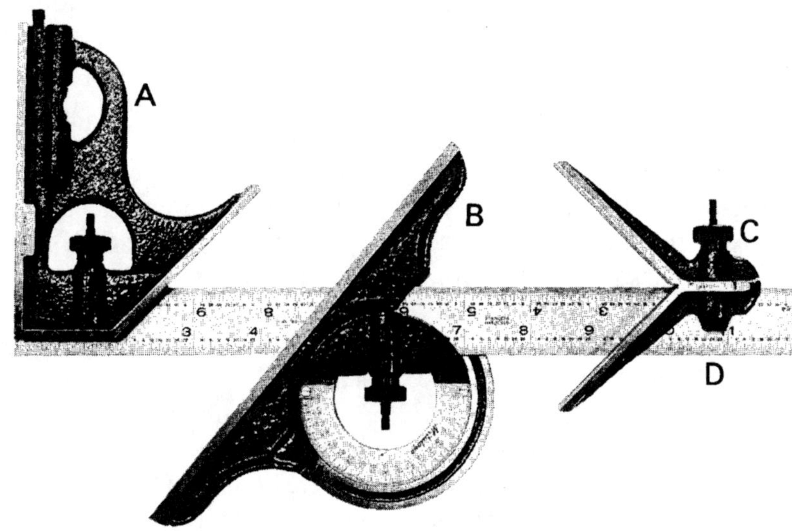

Fig. 9-25 Combination square: (A) Square head, (B) protractor head, (C) center head, and (D) steel rule (MTI Corporation)

position. The head can be swiveled and locked at any angle.

How To Use The Center Head

1. Remove all burrs and apply layout fluid to the work end to make the layout lines more visible.
2. Place the center-head V opening against the work and have the blade adjusted to extend across the flat end of the round work.
3. Scribe a line along the inner edge of the rule.
4. Move the center head 1/4 turn and scribe another line.
5. Center punch at the point the two lines cross and check the accuracy.

OTHER SEMIPRECISION LAYOUT TOOLS

Straightedges, Figure 9-27, are made of hardened steel with graduated and plain edges. Some have beveled edges for easier line visibility. They are used to check parts for

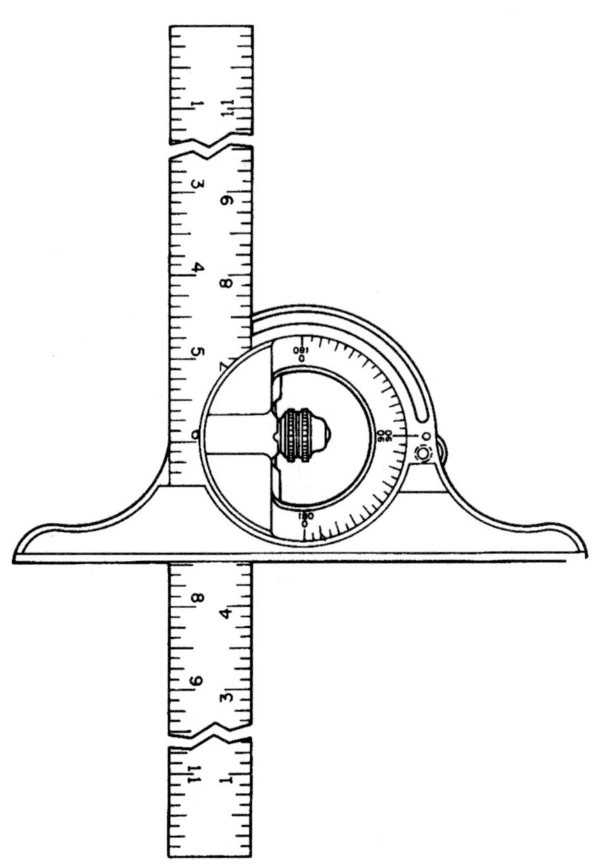

Fig. 9-26 Bevel protractor (L. S. Starrett Company)

Fig. 9-27 Straightedge (L. S. Starrett Company)

straightness. Other uses are to check the trueness of one part to another, Figure 9-28, and to scribe longer, straight lines on parts.

The *double square* is shown in Figure 9-29. It is used for the layout of angles and to check drill and countersink angles. The *steel protractor*, Figure 9-30, is a highly-useful, accurate tool for setting bevels, transferring angles, as a small T square, checking cutter clearance, and many other similar applications.

The *bench block* is used to hold small work when driving pins or drilling, Figure 9-31. A V groove across the face allows round work and odd-shaped stock to be positioned. The holes are made standard size to align a tap square with the work surface to be tapped. Further detail is given in the Unit on Taps and Dies. *V blocks* are for general shop use in layout, Figure 9-32. The blocks are made of cast iron or steel in a wide range of sizes. Some V blocks are provided with U-shaped clamps to firmly hold round work during layout, drilling, milling, or grinding operations.

The *bench plate* is a cast iron or steel plate machined to a flat surface varying in size, Figure 9-33. The plate is used as a base to do general layout work that does not require the accuracy of a surface plate.

The *surface gage* is a tool used for scribing lines at a given height on a workpiece from some face of the work, Figure 9-34. This tool is used on a surface plate or other true base surface.

A groove in the base of the surface gage permits it to be used on cylindrical work, Figure 9-35A. The gage is also used as a precision height gage, Figure 9-35B, and for checking the level of work mounted in a machine vise. The surface gage is also used to transfer lines around several surfaces on the workpiece, lay out duplicate parts to a specific dimension, and to scribe lines on horizontal surfaces.

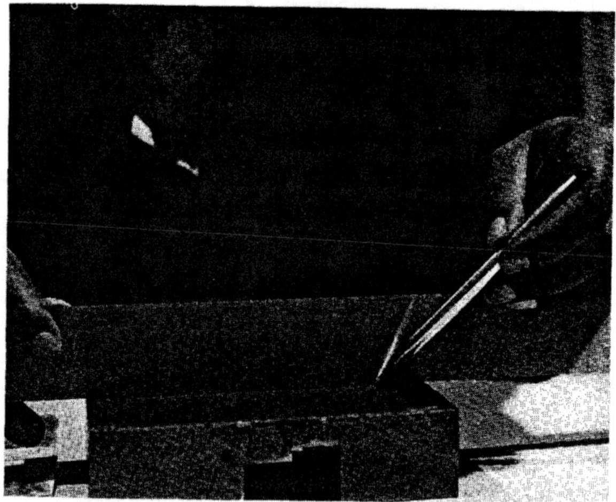

Fig. 9-28 Straightedge used to check flatness of the part surface

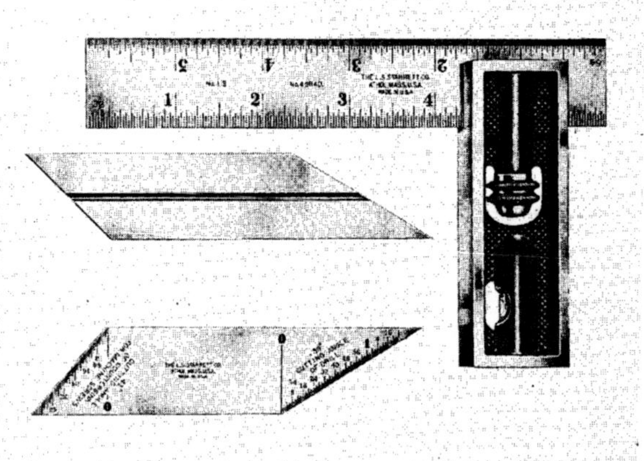

Fig. 9-29 Double square (L. S. Starrett Company)

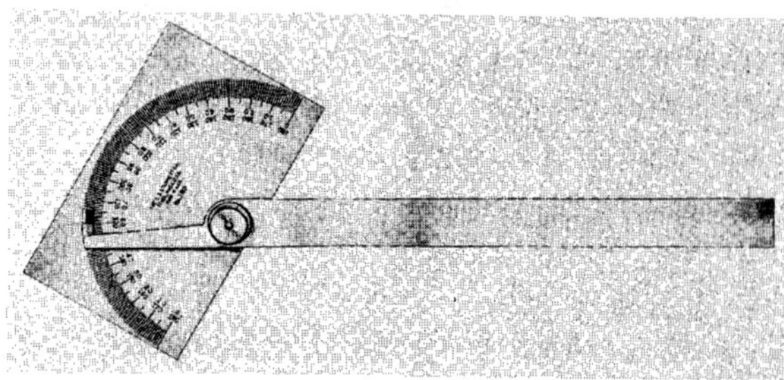

Fig. 9-30 Steel protractor (L. S. Starrett Company)

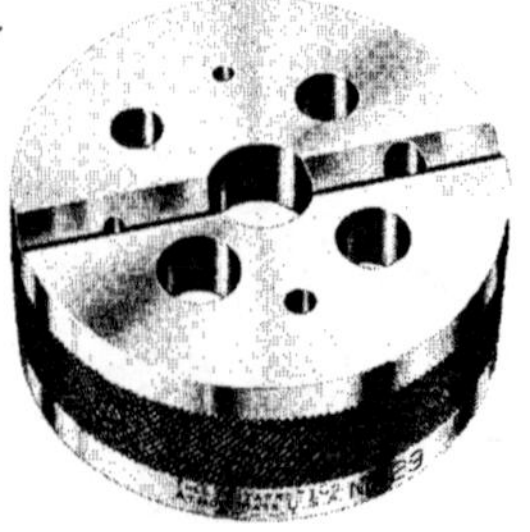

Fig. 9-31 Bench block (L. S. Starrett Company)

Fig. 9-32 V Blocks and clamps

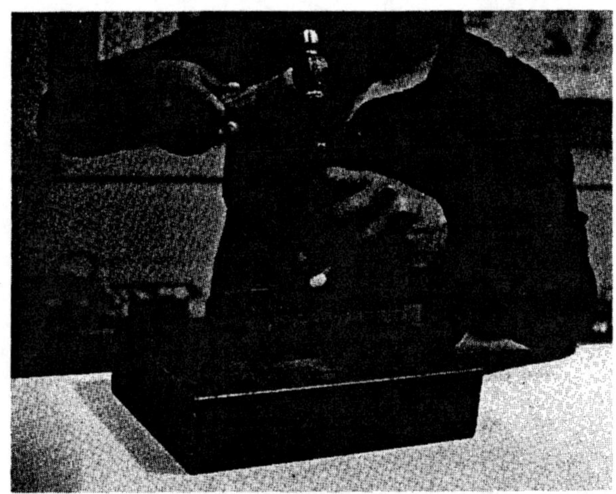

Fig. 9-33 The bench plate is used to support work while center punching the layout.

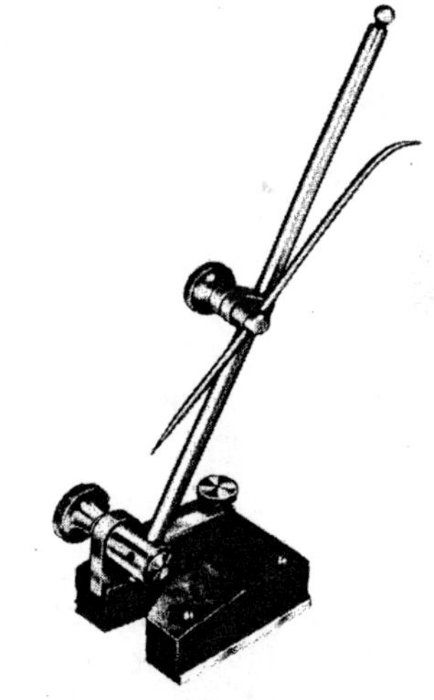

Fig. 9-34 Surface gage (L. S. Starrett Company)

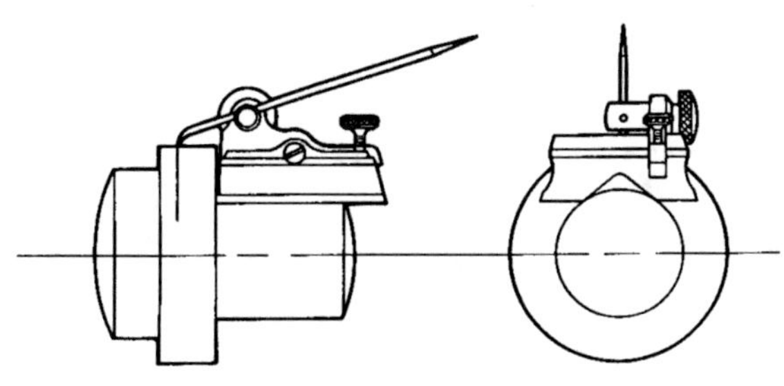

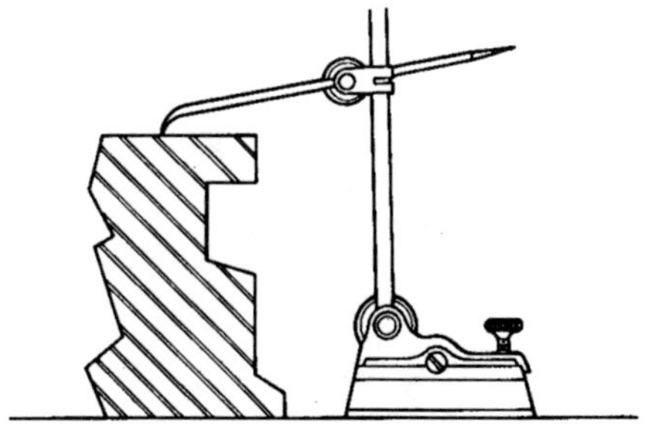

A

B

Fig. 9-35

REVIEW QUESTIONS

A.   Multiple Choice

1. The scriber is held in a manner similar to a:
   a. prick punch.   c. center punch.
   b. pencil.   d. layout hammer.

2. Which of the following is used to sharpen the points of dividers?
   a. Bench grinder   c. Oilstone
   b. Pedestal grinder   d. File

3. Prick and center punches are kept sharp by using a(n):
   a. knife.   c. file.
   b. bench grinder.   d. oilstone.

4. Which of the following tools is used to scribe lines parallel from an internal or external edge by reversing the bent leg?
   a. Beam trammels   c. Dividers
   b. Hermaphrodite caliper   d. A combination set

B.   Short Answer

5. State the two cautions to follow when using sharp, pointed layout tools.

6. Name the tool used for scribing large arcs and circles.

7. List three uses of the straightedge.

8. The ____________ punch is used to make the first small punch marks on layout lines or at center points where lines cross one another.

9. The ____________ ____________ has an included point angle of 60 degrees to 90 degrees.

10. Name the four main parts of the combination square set. List a use for each part.

ACTIVITY

1. From a sketch provided by your instructor, select the proper semiprecision layout tools to use for the layout. Make the layout on a plate provided by your instructor.

UNIT 10 PRECISION LAYOUT TOOLS

OBJECTIVES

After completing this unit, the student will be able to

- define precision layout.

- identify precision layout tools.

- demonstrate the proper use and care of precision layout tools.

INTRODUCTION

Precision layout tools must be able to provide accurate layout and inspection of parts to ± 0.001 inch (± 0.0254 mm). Precision layout tools are costly. Therefore, they must be correctly used and properly cared for to prolong accuracy and tool life.

As with measuring tools, most all layout tools are made with metric dimensions.

Fig. 10-1 Surface plate made of block granite (The Challenge Machinery Company)

CARE OF PRECISION LAYOUT TOOLS

- When stored, give the tools a light coating of machine oil. This prevents rust and corrosion.

- Special hardwood boxes are usually provided for safe, dry storage.

- All layout tools should be stored separately to protect the precision surfaces, edges, and points.

PRECISION LAYOUT TOOLS

Surface Plate

The *surface plate* is a precision, flat-surfaced device made from good grades of cast iron, granite, or ceramic materials, Figure 10-1. Most surface plates are smoothed flat to within very fine tolerances. Surface plates are used where accuracy is required in layout and inspection work.

Make sure surface plates are properly cared for and correctly used. These are important factors in protecting their dependable precision accuracy. Surface plates must not be used as an anvil. Hammering or dropping heavy objects on the plates must never be done.

Steel Parallels

Steel parallels are 4-sided precision bars (rectangular or square), Figure 10-2. They are made of special tool steel, hardened, and accurately ground to standard sizes. The bars are usually made in 6-inch lengths and available in pairs or in complete sets in finished cases. Parallels are used in layout work to raise the work to a suitable height and provide a solid, accurate seat. When used under work on the surface plate, the work surface is held parallel to the top of the surface plate.

How to Use Steel Parallels

1. Remove the burrs from the work.
2. Clean the base surface.
3. Place the parallels in position.
4. Set the work on the parallels.

Angle Plate

The *angle plate* is a precision, L-shaped piece of steel or cast iron accurately machined to a 90-degree angle, Figure 10-3. All edges of an angle plate are machined true and square with each other. Work clamped to an angle plate is thus held parallel and at right angles to the surface plate. Work can be clamped to the angle plate, and all horizontal lines can be laid out. Without removing the clamps, the angle plate is laid on an edge, and all vertical lines are laid out, Figure 10-3A. The horizontal and vertical lines are at right angles to each other.

Steel Square

The *solid or precision hardened steel square*, Figure 10-4, is the most accurate square made. The two parts of the square are the beam and the blade. Both parts are

Fig. 10-2 Magnetic steel parallels (MTI Corporation)

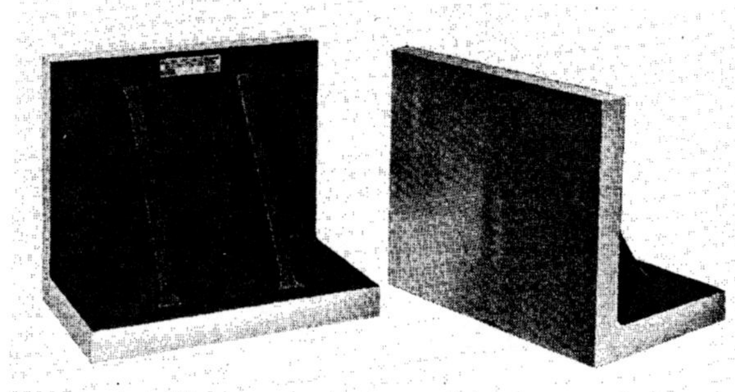

Fig. 10-3 Precision angle plates — all sides and surfaces are at a 90-degree angle to one another. (The Challenge Machinery Company)

Fig. 10-3A Angle plate laid on its side to lay out the vertical lines

Fig. 10-4 Precision, hardened steel square

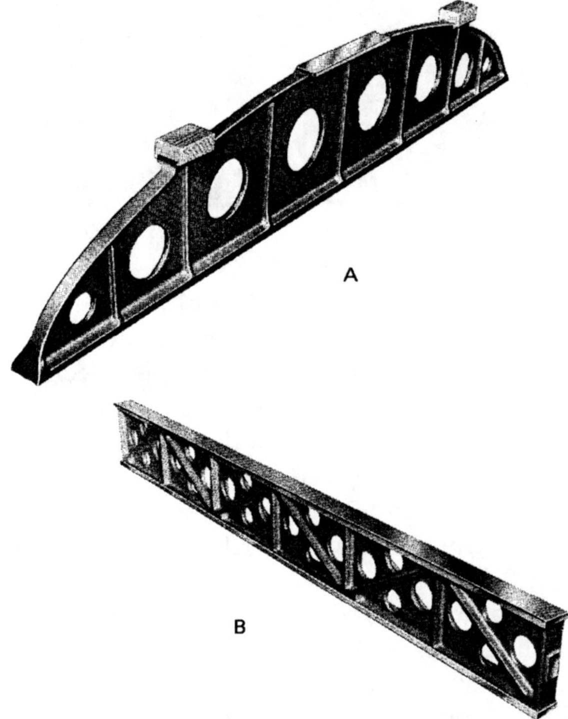

Fig. 10-5 (A) Precision parallel, and (B) precision parallel straightedge (The Challenge Machinery Company)

Fig. 10-6 Precision V blocks (magnetic) (The Challenge Machinery Company)

hardened, tempered, ground, and smoothed for lasting accuracy. This square is recommended for use where extreme accuracy is required. This includes:

- Checking surface flatness.
- Inspecting and checking square precision work on surface plates and machine tools.
- Checking if one surface is square with another.
- Laying out lines at right angles to a machined surface.

Note: The beam is grooved at the inner corner to allow clearance for checking the accuracy of a part. Never use the square on rough work.

V Blocks And Straightedges

Precision straightedges are shown in Figures 10-5A and B. They are used to check the accuracy of long, flat, surfaces and to lay out long, straight lines.

Precision V blocks are shown in Figure 10-6. They are used to hold round work for layout and inspection, Figure 10-7.

Vernier Height Gage

The *Vernier height gage* is a Vernier caliper mounted vertically on a suitable base, Figure 10-8. The upright bar and movable jaw are similar to the standard Vernier caliper. The movable jaw has a hardened scriber to scribe lines accurately to within 0.001 inch

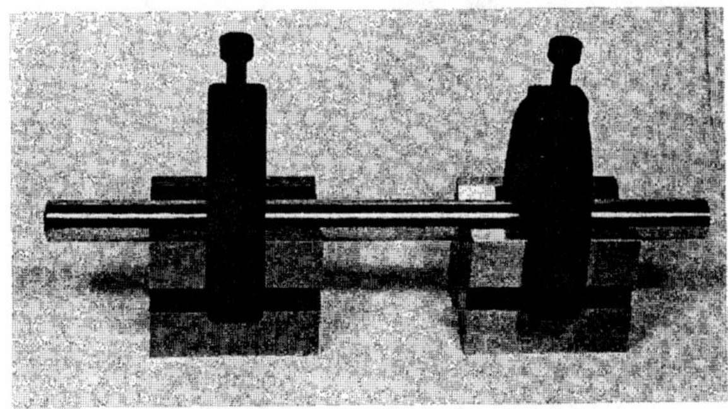

Fig. 10-7 V blocks holding a round part
(MTI Corporation)

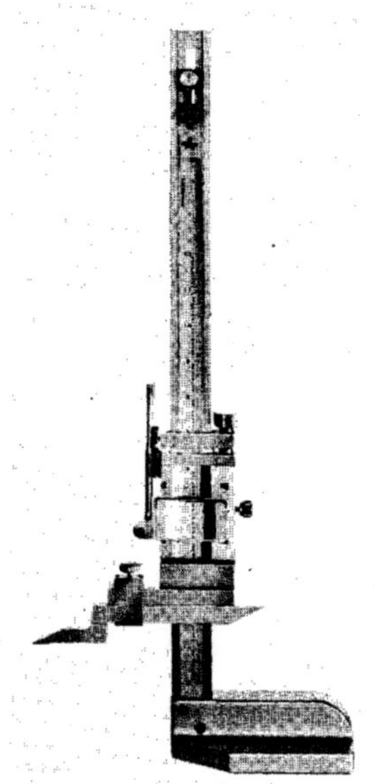

Fig. 10-8 Vernier height gage (MTI Corporation)

(0.0254 mm). This tool is used with the surface plate where more accurate layout lines are required, Figure 10-9. Other uses of the Vernier height gage are shown in Figure 10-10A and B.

Universal Bevel Protractor

The *Universal bevel protractor* uses a Vernier to accurately measure parts of a

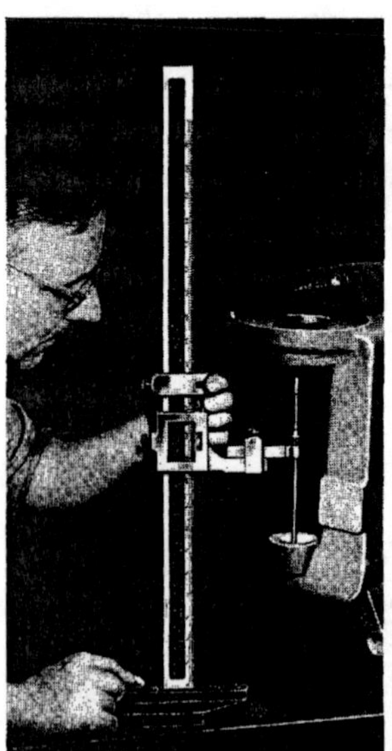

Fig. 10-10 (A) Use of the Vernier height gage to measure a depth, and (B) use of the same gage to measure a machined surface (L. S. Starett Company)

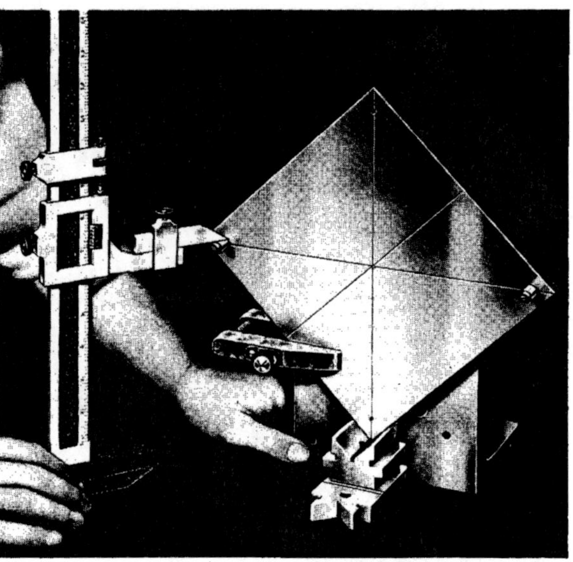

Fig. 10-9 Accurate use of the Vernier height gage
(L. S. Starrett Company)

circle, Figure 10-11. This tool is used in the same manner as the bevel protractor of the combination set except for greater precision.

Angular measurements can be read to 5 minutes (5′) or 1/12 of a degree. The dial of the protractor is graduated to the right and left of 0 up to 90 degrees. The Vernier scale is also graduated to the right and left of 0 up to 60 minutes (60′), each of the 12 Vernier graduations representing 5 minutes. Since both the protractor dial and Vernier scale have graduations in opposite directions, any size angle can be measured. The Vernier

Fig. 10-11 Universal bevel protractor
(MTI Corporation)

Fig. 10-12 Reading a Vernier
on a bevel protractor

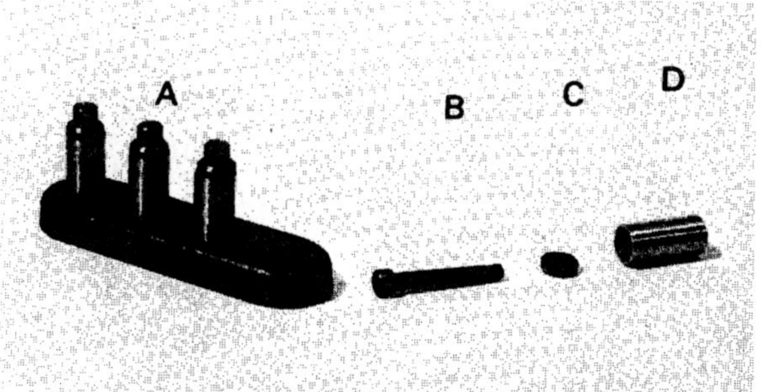

Fig. 10-13 Toolmaker's buttons: (A) Holder, (B) screen, (C) washer, and (D) precision hardened sleeve (L. S. Starrett Company)

reading must be read in the same direction from 0 as the protractor, either left or right.

On the Vernier scale 12 graduations occupy the same space as 23 graduations or degrees on the protractor dial. Therefore, each Vernier graduation is 1/12 degree or 5 minutes shorter than 2 graduations on the protractor dial. When the 0 graduation on the Vernier scale coincides with a graduation on the protractor dial, the reading is in exact degrees. Should some other graduation on the Vernier scale coincide with a protractor graduation, the number of Vernier graduations is multiplied by 5 minutes. This amount is added to the number of degrees read between the zeros on the protractor dial and Vernier scale.

Reading the Universal Bevel Protractor

To read the Universal bevel protractor with a Vernier, Figure 10-12, proceed as follows:

1. The 0 on the Vernier scale lies between 50 and 51 degrees on the protractor dial to the left of the 0°. This indicates 50 whole degrees that the Vernier 0 line has passed.
2. Also, reading to the left, the 4th line on the Vernier scale coincides with a graduation line on the protractor dial as indicated by the stars. Each line on the Vernier represents 5 minutes. Therefore, 4 lines times 5 minutes, or 20 minutes, are to be added to the number of whole

degrees. The total reading of the protractor is 50 degrees and 20 minutes (50°20′).

Toolmaker's Buttons

Toolmaker's buttons, Figure 10-13, are steel bushings which are tightened in exact positions on the work after the holes are laid out. The buttons are hardened and ground to a given diameter in even tenths of an inch. The ends are ground square and seat firmly and squarely over the center of a hole.

Each set consists of four buttons fastened to a steel base plate to protect the screws and washers that are used to clamp them. One button in each set is 1/8 inch longer than the others to allow truing up when 2 buttons are used closely together. The holes in the buttons are made larger than the screws to permit sidewise adjustment after the button is fastened to the work.

How to Use Toolmaker's Buttons

1. The work to which the buttons are to be clamped must be true and parallel.
2. Lay out all holes to be drilled or bored in the workpiece.
3. Prick punch the intersections of points to be drilled or bored.
4. Drill and tap for the button screw to enable the button to be clamped to the workpiece.
5. Remove all burrs caused by drilling and tapping.
6. Tighten the buttons in position. Tighten them enough to allow them to be

positioned by lightly tapping with a soft-faced hammer.

7. The buttons are then accurately set in position with the use of a micrometer or other precision measuring tool.

8. After setting, the buttons are tightened in place and rechecked for accuracy.

9. The workpiece is then ready to be mounted on the proper machine tool for the required machining operations.

Many other layout tools are available from suppliers that provide super precision layout and inspection. Precision layout requires more accuracy than does semiprecision layout, although many semiprecision tools are used in precision layout. Greater care must be taken to think and work to these closer tolerances. It is important to select only those layout tools and procedures to meet the part tolerances.

REVIEW QUESTIONS

A. Multiple Choice

1. Which of the following layout tools is the most accurate to use to check a part for squareness?
   a. Combination square
   b. Solid or precision square
   c. Double square
   d. Steel protractor

2. Which of the following states the correct storage of precision layout tools?
   a. In the tool tray with other tools
   b. Hanging on the wall
   c. Separately in special hardwood boxes
   d. On the surface plate

3. The surface plate may be used:
   a. as an anvil.
   b. as a layout bench.
   c. as a bench plate.
   d. only for precision layout and inspection.

4. Steel parallels are generally used with a(n):
   a. surface plate.
   b. rough bench plate.
   c. work bench.
   d. angle iron.

5. Which of the following layout tools is best to measure an angle surface?
   a. Toolmaker's buttons
   b. Vernier height gage
   c. Bevel protractor
   d. Steel square

B. Short Answer

6. The Universal bevel protractor uses a ___________ to accurately measure the parts of a circle.

7. What tool is used to hold round work for layout and inspection?

8. ___________ are used in layout work to raise the work to a suitable height and provide a solid, accurate seat.

ACTIVITY

1. With precision tools provided by your instructor, demonstrate the correct use of each.

UNIT 11 LAYOUT PROCEDURES

OBJECTIVES

After completing this unit, the student will be able to

- define the necessity of layout as applied to shop practices.
- describe layout procedures from print to workpiece.

INTRODUCTION

The benchworker is frequently involved in layout work. This process involves the accurate transfer of measurements from a shop print to the workpiece. The shop prints are made by drawing lines on paper (drafting). Layout work is the scribing (scratching) of lines on the workpiece surface.

These scribed lines accurately locate centers for holes and circles. They also define boundaries of shape, size, and other locations. Layout lines serve to guide the worker in making the part to the given dimensions. A layout is comparable to a single view (end, top, or side) of a part which is scribed directly on the workpiece, Figure 11-1.

In layout work, the benchworker must know how to carefully read prints and accurately transfer print dimensions to the workpiece. Selection and use of the proper layout tools are equally important. After careful study of the print, the correct materials are selected and laid out to the degree of accuracy called for on the print.

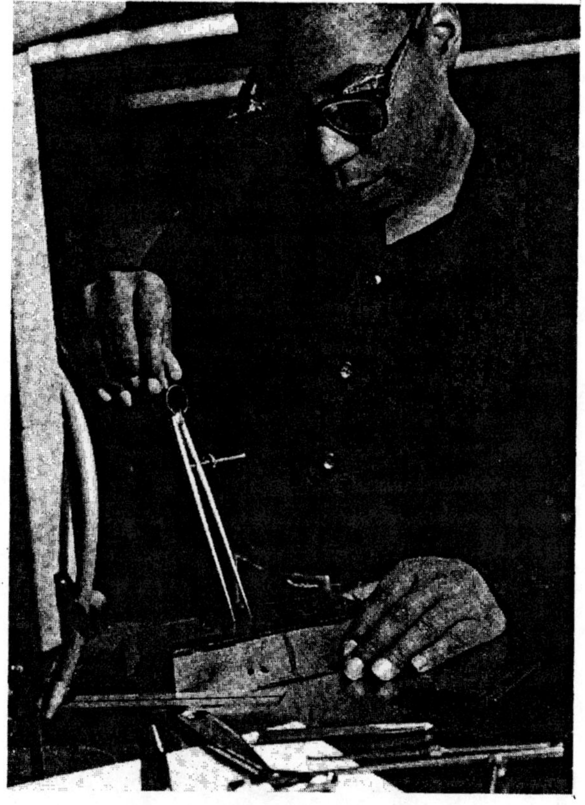

Fig. 11-1 Typical layout on a plate

PRINT CARE

- Be careful when handling all blueprints and working drawings. Keep the drawings clean and in order.

- When the job is done, return blueprints to the proper file or storage area.

- Make sure blueprints and drawings are not mutilated, destroyed, or lost.

LAYOUT GROUPS

Layout work may be divided into two general groups: semiprecision layout accuracy of ± 1/64 inch, (± 0.4 mm), and precision layout accuracy of ± 0.001 inch (± 0.0254 mm). These tolerances depend on the part requirements and print specifications.

LAYOUT PROCEDURES

Layout work must be done with tools less convenient than those used for drafting. Layouts are generally required on parts that require hand or visual alignment to the cutting tool. Layout is often used in drill press work. Through proper layout, the worker can more clearly check the steps of certain operations. Thus, workpieces will be completed in an efficient and orderly manner.

Surface Preparations

A layout fluid is used to provide a contrasting background on the work surface. This coating shows the scribed line clearly. Chalk may be rubbed into a rough casting to make layout lines more visible. Marking inks such as Dykem Blue and Starrett Layout Fluid are typical layout fluids used in the shop. Some parts may be colored by heat to a blue color. This surface color allows scribed layout lines to be clearly seen.

Reference Points

In layout work, a baseline must be selected from which to begin all layout dimensions. This serves as a starting point from which to measure dimensions, angles, and lines of the layout, Figure 11-2.

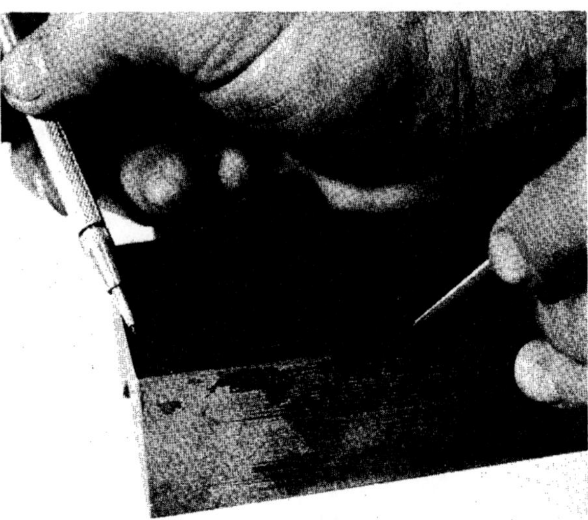

Fig. 11-2 Established horizontal and vertical baselines

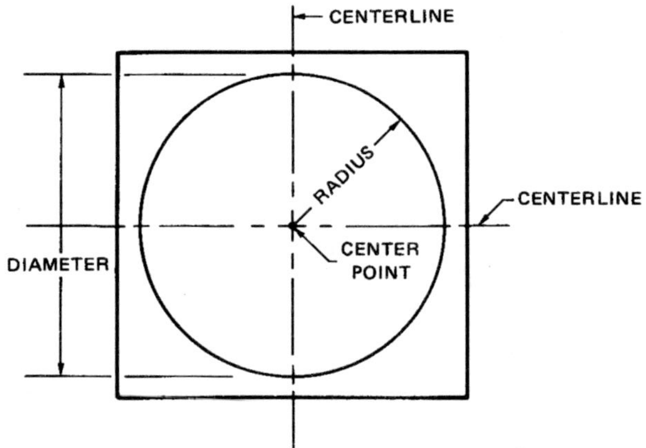

Fig. 11-3 Circular layout from center point and radius or diameter

A machined edge or a scribed line or centerline may be used as the reference or baseline. This reference point leads to accuracy when making layouts because errors are limited between the reference line and one specific line or point. Circular parts, such as flanges, are usually laid out from a center point and a diameter or radius line, Figure 11-3.

Accepted Layout Procedure

1. Establish a horizontal and vertical baseline or reference line.

2. Make each layout line or point from these baselines. Making a layout with each

line or point measured from the preceding line or point causes compounding of previous error and results in an inaccurate layout.

3. Use only clean, sharp scribed lines. Any double or sloppy lines should be removed by cleaning or applying another coat of layout fluid and the line rescribed.

4. A scribed line on the surface of plain metal is usually difficult to see. Any hot-rolled material with surface scale should have the scale removed before laying out the surface. The scale does not allow for clean, sharp layout lines.

HOW TO MAKE A LAYOUT

The job is made easier, completed quicker and with fewer errors by making and using a logical work plan. Use the following step-by-step procedure to develop a work plan for making a layout.

1. Study the blueprints or working drawings carefully.
2. Select the correct kind and size of material described in the bill of materials.
3. Cut material to correct size, length, and quantity. Be sure to return uncut material to its proper storage.
4. Remove all sharp edges and burrs. Use a grinder or a file with a handle to remove burrs. Burrs can cause inaccurate layout and personal injury.
5. Clean the workpiece of all grease and oil.
6. Apply the proper surface preparation to clearly show layout lines.
7. Select proper layout tools and keep the tools organized.
8. Establish and mark accurate reference edges to work from.
9. Mark the lower left-hand corner X and work from edges A and B, Figure 11-4. If edge A is not true, then scribe a vertical line 90 degrees to edge B.
10. From edges A and B, lay out all horizontal and vertical centerlines to locate hole locations. Always lay out all lines from the baseline reference.

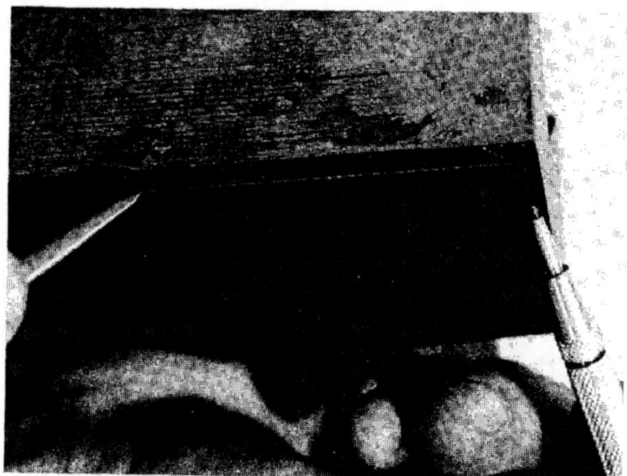

Fig. 11-4 The lower, left-hand corner is marked with an X. Work from the reference edges A and B.

11. Check center-to-center layout of centerlines with print and make any corrections.
12. Scribe all arcs and circles and lay out all angular lines and point-to-point connecting lines.
13. Use the prick punch to mark the points where the horizontal and vertical centerlines intersect.
14. Make sure the punched mark is exactly at the cross of the centerlines. For greater accuracy, use a magnifying glass.
15. Scribe reference lines to show the diameter location of holes to be drilled.
16. Lightly prick punch the circumference of these scribed reference circles. The prick punches will remain even if the scribed circles are scratched away by chips when drilling. These punch marks provide indicators to determine the proper location of the drill bit as the hole is started.
17. Center punch all center points. This larger indent provides a good starting place for the twist drill to start.

Note: When dimensions are given on the print in inches or fractions, accuracy of ± 1/64 inch (± 0.4 mm), is accepted.

CASTING LAYOUT

Uneven castings require skill in judgment and calculation to properly lay out the part so it may be finished to print specifications. An incorrectly laid out casting may not finish to size. Use the following steps as a guide:

1. Measure the overall casting, Figure 11-5.
2. Establish horizontal and vertical reference centerlines, Figure 11-6.
3. Lay out the outline lines from the centerlines.
4. Be sure enough material shows beyond the layout lines to produce the desired part.
5. Fill the center of the hole of the casting with a piece of tin to provide for the centerline, Figure 11-7.
6. Scribe the cross centerlines and prick punch the center of the intersecting lines, Figure 11-8.
7. Scribe the arc to lay out the hole pattern, Figure 11-9.
8. Center punch the intersection of the lines to locate the holes to be drilled, Figure 11-10.

LAYOUT OF MACHINED PARTS

Machined parts and finished surfaces can be laid out to a finer degree of accuracy than rough castings or unfinished work. Some parts require a secondary layout. The part may need to be laid out, machined, then laid out again to machine to new guidelines. Do not make layout lines which need to be removed and laid out again. This is a waste of time.

How to Find the Center of Open-Center Parts

Some parts have no center material to mark upon. A flange for the end of a pipe is an example of a part with an open center, Figure 11-11. If a center point must be established to scribe a bolt circle, follow the steps below.

1. A piece of metal or other suitable material is wedged into the opening to bridge the hole, Figure 11-12.
2. Apply the layout fluid, Figure 11-13.
3. By referring to a table of *chords* (distance between centers of holes around a circle),

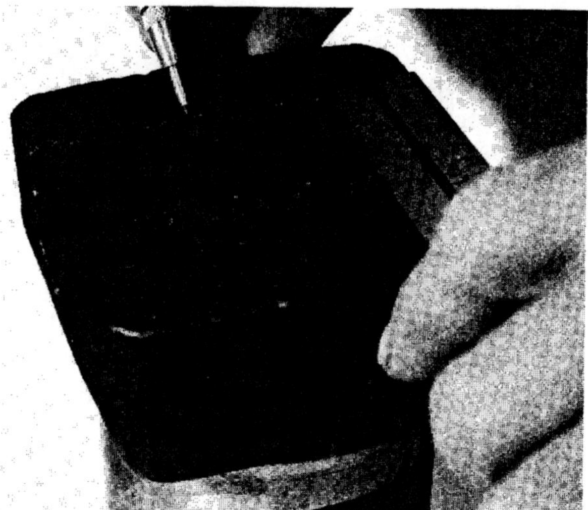

Fig. 11-5 Measure the overall casting.

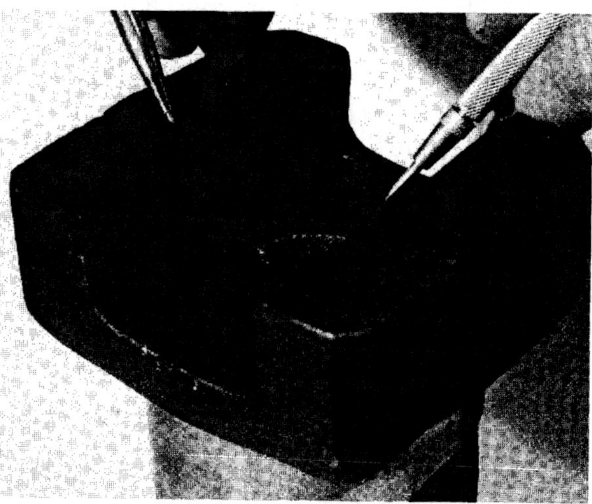

Fig. 11-6 Establish center layout lines.

Fig. 11-7

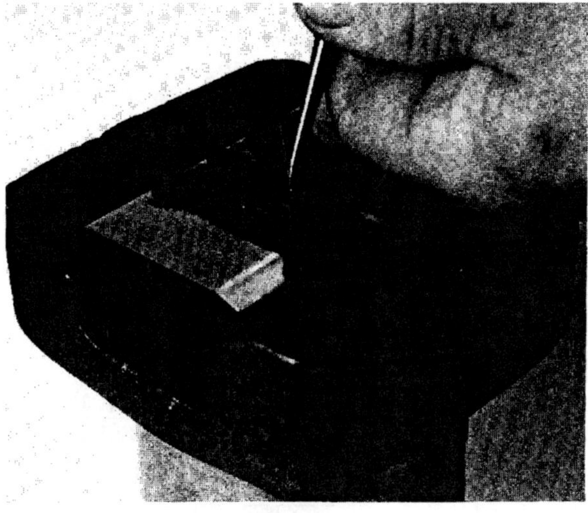

Fig. 11-8

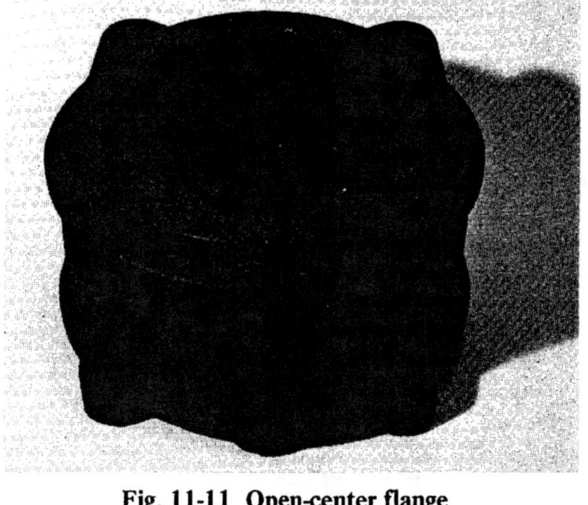

Fig. 11-11 Open-center flange

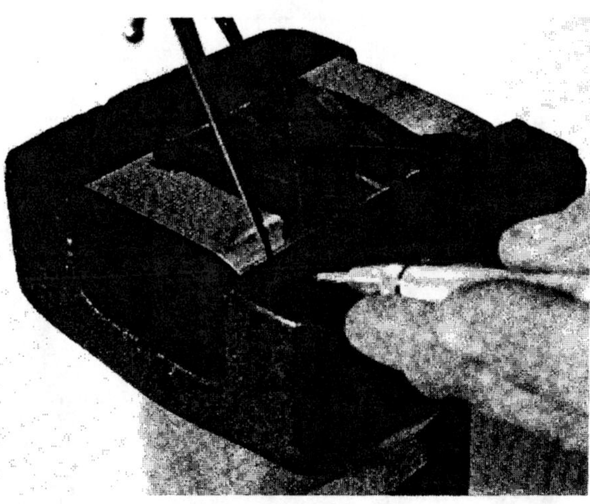

Fig. 11-9

Fig. 11-12 Piece of tin to fill the
open hole to provide a center for the layout

Fig. 11-10

Fig. 11-13 Applying layout fluid

Number of Divisions in the Circle	*Length of Chord
3	0.866
4	0.707
5	0.588
6	0.500
7	0.434
8	0.383
9	0.342
10	0.309
11	0.282
12	0.259
13	0.239
14	0.223
15	0.208
16	0.195
17	0.184
18	0.174
19	0.164
20	0.156
21	0.149

Number of Divisions in the Circle	*Length of Chord
22	0.142
23	0.136
24	0.131
25	0.125
26	0.121
27	0.116
28	0.112
29	0.108
30	0.105
31	0.101
32	0.098
33	0.095
34	0.092
35	0.090
36	0.087
37	0.085
38	0.083
39	0.080
40	0.078

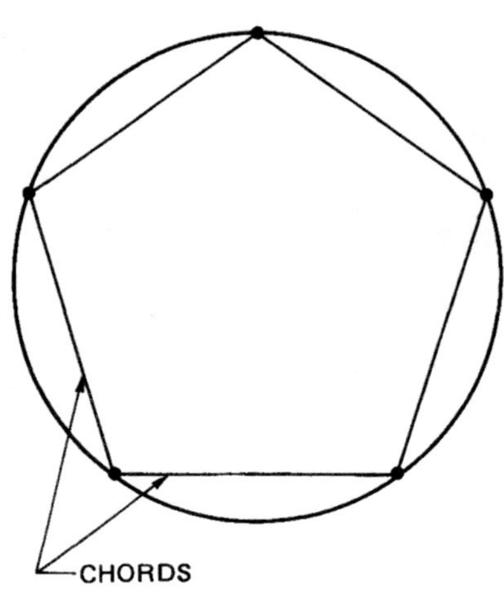

*Note: This table is calculated for circles having a diameter equal to one-inch diameter. For other circles, multiply the length given by the diameter of the circle.

Fig. 11-14 Table of chords

the layout of a number of equally spaced holes is easily done, Figure 11-14. These tables provide constants for various numbers of spaces which are simply multiplied by the diameter of the circle.

4. The result is the distance to set the dividers or trammels to scribe the center distances of the holes, Figure 11-15.

To find the center of circular, solid-center parts, refer to the section on hermaphrodite calipers in Unit 9.

LAYOUT FOR MULTIPLE SAME PARTS

All layout work should be done as quickly, accurately, and simply as the workpiece allows. When multiple parts are to be made requiring the same layout, use the following steps as a guide:

1. Before beginning the layout, be sure you have the correct print for the part to be made or repaired.

Fig. 11-15 Punch the center and make the scribed arc for the hole locations.

2. One part is accurately laid out.

3. This layout is checked for accuracy and the part machined.

Fig. 11-16 Templet clamped to a part to be drilled in the same hole locations

4. After machining, the part is checked with the print. If the part is correct, it is used as a templet or pattern to make the remainder of the parts.

5. The templet is clamped to the unfinished parts one at a time, and the machining operations are performed, Figure 11-16.

Only one part needs to be laid out, therefore saving valuable production time. Chances of error in many layouts of the same part are minimized.

Note: Never take a measurement with a rule directly from the print because of scaling errors. *Scaling* refers to the scales used in drawing, such as $1/4'' = 1''$. Print paper also stretches and shrinks with changes in weather conditions. Dimensions must be taken only from the figures shown with the dimension lines.

REVIEW QUESTIONS

A.   Multiple Choice

1. Correct layout work is important to maintain:
   - a. workpiece accuracy.
   - b. reduced scrap parts.
   - c. interchangeability of parts.
   - d. all of the above.

2. Which of the following statements best describe *layout*?
   - a. Measuring from a print and transferring these measurements to the workpiece
   - b. Measuring the workpiece and comparing this measurement to the print
   - c. Scribing of lines on the workpiece surface
   - d. Making a detail drawing on the workpiece.

3. When laying out a casting, which of the following is done first?
   - a. Establish reference centerlines.
   - b. Fill the center of the casting with a piece of tin to provide for centerlines.
   - c. Measure the overall casting.
   - d. Layout the outline lines from the centerlines.

4. Which of the following surfaces can be laid out to a finer degree of accuracy?
   - a. Rough surfaces
   - b. Machined and finished surfaces
   - c. Square surfaces
   - d. Round surfaces

5. Which of the following tools is used to transfer machining locations when making multiple parts?

    a. Beam trammels   c. Hermaphrodite calipers

    b. Dividers   d. Templets

B. Short Answer

6. Why is layout fluid used on the work surface before scribing layout lines?

7. Why is it considered bad practice to lay out lines on a material having surface scale?

8. List the two general groups in which layout work may be divided. State the tolerances for each.

ACTIVITY

1. From the following sketch, arrange the list of layout procedures in their proper order by indicating numbers 1 through 14.

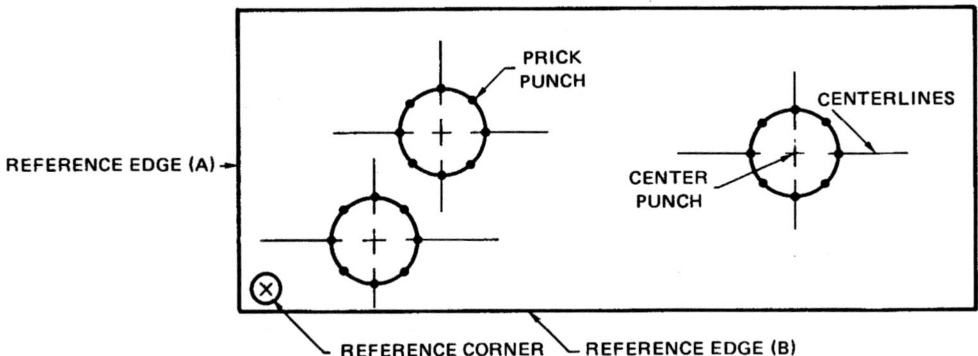

a. __________ Center punch the center points.

b. __________ Check the layout and correct any errors.

c. __________ Establish the layout reference edges and corner.

d. __________ Study the print carefully.

e. __________ Scribe the print carefully.

f. __________ Cut the material to correct size, length, and number of pieces.

g. __________ Prick punch the circumference of the reference circles.

h. __________ Layout and scribe the centerlines to establish hole locations. Measure from edge A and B and centerline to centerline.

i. __________ Remove all the sharp edges and burrs from workpiece.

j. __________ Select and properly organize the layout tools needed.

k. __________ Apply a thin coating of layout solution over surface to be laid out.

l. __________ Prick punch the center points at the cross of the vertical and horizontal centerlines.

m. __________ Select the proper kind and correct shape material.

n. __________ Clean the surface of all loose surface scale, dirt, and grease.

SECTION IV
HAND TOOLS

UNIT 12 ASSEMBLY TOOLS, ARBOR AND SHOP PRESSES

OBJECTIVES

After completing this unit, the student will be able to

- identify the common noncutting hand tools used by the benchworker.
- explain the importance of using the proper tool for a given job.
- define the correct use and care of a selected tool.

INTRODUCTION

Hand tools are those tools which are powered by hand. These tools become an extension of the worker's hands. They provide the needed turning power or torque (rotation) in the fitting of parts and fasteners. Difficult tasks can be done quickly and easily with the use of the proper hand tools. The worker must be skillful in the identification, selection, and use of these tools to complete work safely and accurately. Use hand tools in the manner for which they were designed. This reduces damage to expensive tools and workpieces. Proper tool usage also reduces personal injury.

> **Caution:** Do not use metric tools on English fasteners or English tools on metric fasteners. They do not fit. Injury and damage can result.

GENERAL CARE OF HAND TOOLS

- Place the tools that are being used within an easy reach in an organized, safe place.

- Tools not in actual use must also be protected so they do not fall to the floor and become damaged or broken.

- Clean, sharpen, and return tools to their proper storage place when finished using them.

- Keep tools properly lubricated and protected from rust.

- Protect tools with sharp edges from touching one another.

- Keep burrs removed from tools.

- Use the proper tool designed for a job.

VISES

The *bench vise*, or *machinist's vise*, Figures 12-1A and 12-1B, is used to hold work while it is being assembled, disassembled, or otherwise worked on at the bench. The vise becomes a third hand to the worker. The vise also rigidly holds the workpiece while various machining operations are being done. Work held in a vise can be more safely and accurately worked on.

The vise is made of either cast iron or steel. One jaw is attached or made solid to a base. The other jaw is adjustable by a handle through a screw. Vises are sized by the width of the jaw, Figure 12-2. Some vises are made with a quick adjusting lever to position and set the jaw against the work. Vises are made with either solid or swivel bases. The solid vise is the most rigid.

The inner-contact jaw faces of a vise are made of grooved, hardened steel to grip rough parts. Finished parts and surfaces must be protected from these jaws. Soft jaws are made to slip over the regular vise jaws. They are made from aluminum angle shapes, copper, or other soft, nonmarking materials, Figure 12-3. Some other uses of the bench vise include gripping work for filing, holding work for bending, and straightening sheet metal on the anvil.

Use and Care of the Bench Vise

- Mount the vise firmly at a convenient working height on the bench.

- Securely lock the swivel base.

- Never hammer on the handle.

- Never use handle extensions.

- Never hammer on the moving parts of the vise. Hammer on the anvil section. Figure 12-4 shows a bench anvil used to hammer and form metal on.

Fig. 12-1A Bench vise (The Ridge Tool Company)

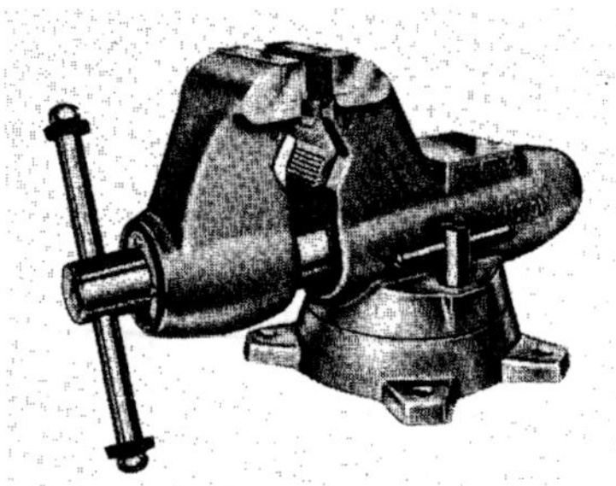

Fig. 12-1B Machinist's bench vise
(Wilton Corporation)

Fig. 12-2 The vise is sized by the width of the jaw.

Fig. 12-3 Soft inserts used to protect finished work from hard, steel-grooved jaws

Fig. 12-4 Bench anvil used to hammer, rivet, and for form metal parts into various shapes

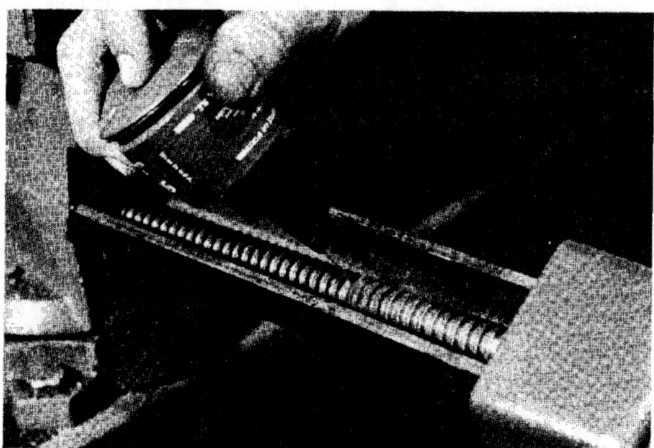

Fig. 12-5 Use dry graphite to lubricate the vise screw and nut assembly

Fig. 12-6 C clamps used to hold one part to another (Wilton Corporation)

- Lubricate the screw and nut assembly with dry graphite, Figure 12-5. It can be applied with a brush or is available in containers.

- Keep jaw faces tight and clean.

- Use the proper size vise for the job.

CLAMPS

Various styles and types of clamps are available to the benchworker to hold parts together for assembly, Figure 12-6. The size of clamps is determined by the largest opening of the jaws. Clamps are sometimes called vises without bases.

Parallel clamps have two screw members, Figure 12-7. The screw near the outside tends to hold the jaws together. The screw on the outside pushes the jaws apart. This leverage clamps the work when the jaws are parallel.

C clamps, or carriage clamps, are shown in Figure 12-6. They are used to hold work to an angle plate, work table, or for clamping two pieces of work together.

Use and Care of Clamps

- Lubricate the screw and nut assembly with dry graphite.

- Keep the swivel foot free-moving and lubricated.

- Store in a clean, dry area when not in use.

- Make sure clamps are never over tightened. This causes damage to the adjusting parts and possibly the workpiece.

Fig. 12-7 Parallel clamps

PLIERS

Pliers are another form of hand pressure clamp. They are sized by their overall length, Figure 12-8.

Slip joint pliers allow the jaws to expand to grasp larger sizes of work. *Tongue and groove pliers* are of the interlocking-joint type, Figure 12-9. The action of these pliers prevents slipping under load. They are used for many purposes.

Other plier styles such as *side cutters* and *diagonal cutters* are used for wire cutting. *Round-nose pliers* are used for looping wire. *Needle-nose pliers*, both straight and bent nose, are used to hold various workpieces.

The *locking plier*, Figure 12-10, is an adjustable lever-jawed, locking-type plier used for high gripping power. A screw in the end adjusts the wrench lever action to the work. Figure 12-11 shows one of the special types often used to clamp parts. The *chain wrench* is used as a clamp or can be used as a wrench to grip and rotate parts.

Snap ring pliers are commonly used to install and remove various retaining snap rings. These pliers are made with replaceable or exchangeable tips. They are designed for both internal and external groove use (Refer to Unit 4, Figure 4-21). .

Many special-shaped pliers are made to aid the benchworker. Refer to a supplier's catalog for a complete listing.

Use and Care of Pliers

- Never use pliers or other wrenches as hammers.

- Occasionally lubricate the moving parts of pliers with dry graphite.

- Keep the gripping jaws sharp by dressing them with an abrasive wheel or file.

HAMMERS

The hammer is probably the most often used of all hand tools. They are made in various styles and sizes, with either hard or soft heads. Hammers are sized by the weight of

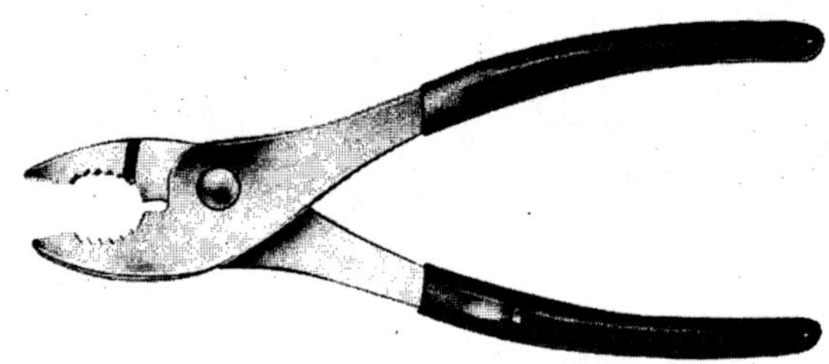

Fig. 12-8 Standard pliers (The Stanley Works)

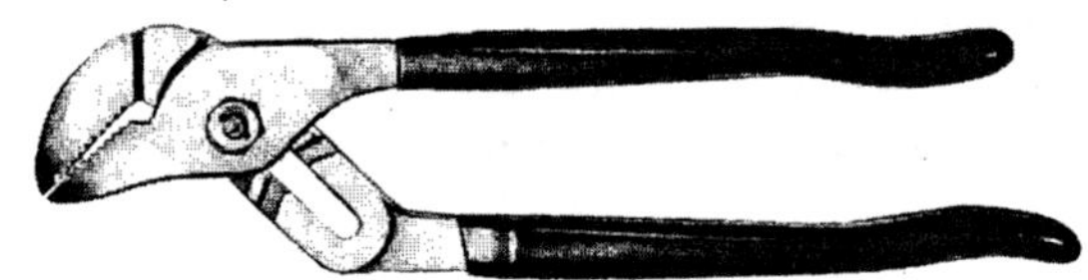

Fig. 12-9 Tongue and groove pliers (The Stanley Works)

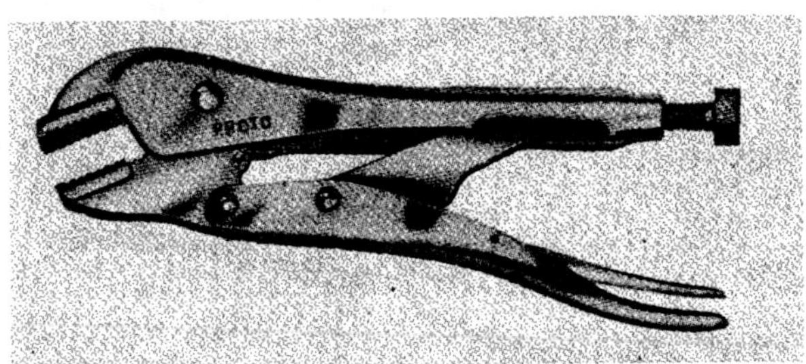

Fig. 12-10 Locking pliers (Proto Professional Tools)

Fig. 12-11 Clamping pliers holding two parts together

Fig. 12-12 Heavy mall or sledge hammer (The Stanley Works)

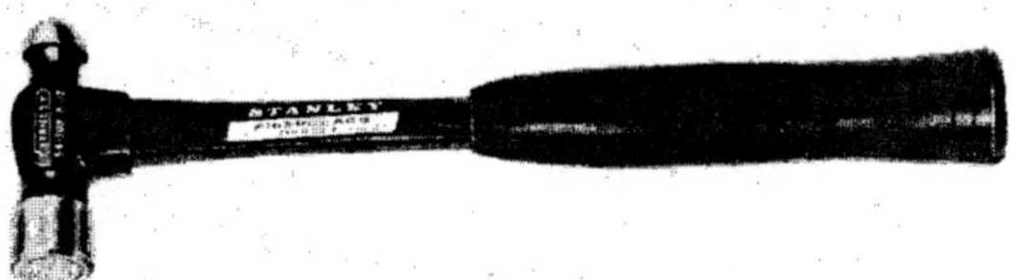

Fig. 12-13 Ball peen hammer (The Stanley Works)

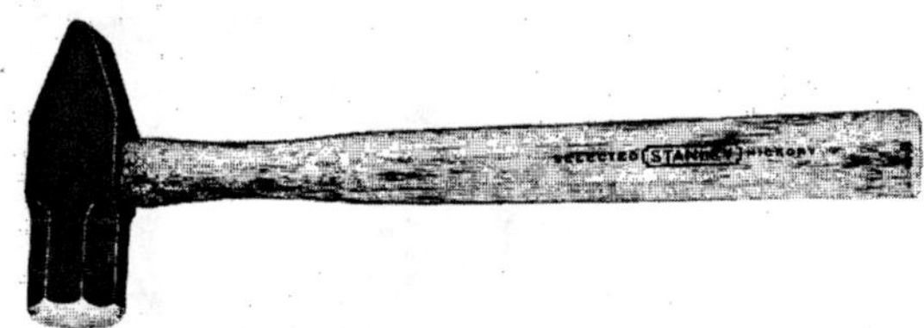

Fig. 12-14 Cross peen hammer (The Stanley Works)

the head. They range from the 2-ounce ball peen to 3-pound malls, Figure 12-12.

Hard hammers, the *ball peen* being the most common, are made of hard alloy steel, Figure 12-13. The *ball-peen rounded end* is used by the benchworker for many uses including the removal of rivets. The other end of the head is used for striking. The *cross peen* and *straight peen hammers* are also used in various types of benchwork, Figure 12-14.

Soft-faced hammers are made with heads of plastic, brass, copper, babbit, rubber, and rawhide. These hammers are used to provide a blow and not damage harder or finished surfaces. The tips of these hammers are replaceable.

Hammer handles are made of good, hard wood or fiberglass stock. They are tapered at the end to prevent the head from slipping down the handle. The head and handle are forced together with a wedge driven into the opening that receives the handle.

Care and Use of Hammers

- Keep the hammerhead tight.
- Never strike one hammer with another.
- Always strike an object squarely with the hammer. Always make sure the face of the hammer is parallel with the surface being struck.
- Always use a smaller hammer for a small job and a sledge or mall for heavier work.
- Grasp the hammer near the end of the handle firmly, but not rigidly. This gives full striking power to the head.
- While striking with the hammer, always watch the part of the workpiece you intend to strike, not the hammerhead.

- Always use proper eye protection when using hammers.
- Never use soft-faced hammers on sharp edges. The corners could damage the hammer.
- Replace cracked or broken handles.
- Remove battered edges on the head by filing or grinding.

WRENCHES

Wrenches are made of heat-treated steel in a wide variety of styles. They are available in both English and metric sizes. Many are alloyed for strength and chrome plated to prevent rusting. Their main function is to hold and turn nuts, bolts, cap screws, plugs, pipes, and various other threaded parts. Standard wrenches are made for practically all operations and services. Many special wrenches are also made. Wrench openings are made slightly larger than the marked size. This is done so the wrench fits easily over the nut or bolt head.

The *adjustable wrench* is a general-purpose, open-end wrench having one fixed jaw which is made a solid part of the wrench. The other jaw is made adjustable to seat snugly to the bolt head or nut. Figure 12-15 shows the correct direction to pull on this wrench.

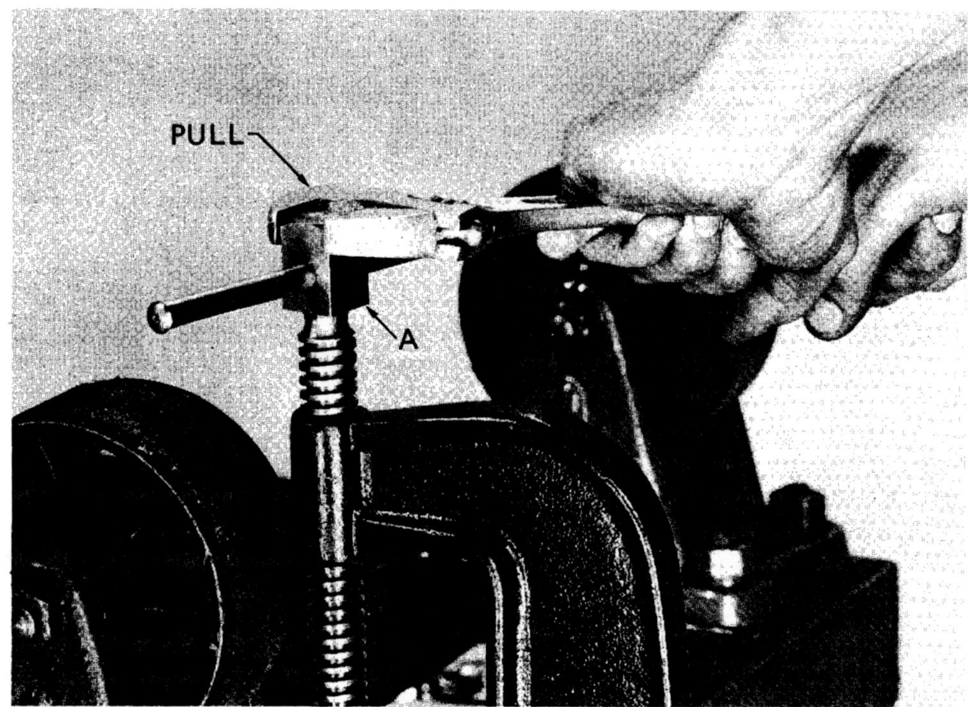

Fig. 12-15 Correct use of an adjustable wrench — pull toward the
movable jaw (A) (Adjustable Clamp Company)

Open-end wrenches usually have a different size on each end, Figure 12-16. The wrench ends are often angled different amounts to be used easily in close areas. *Box wrenches* are also double ended and made offset to clear the user's hand, Figure 12-17. These wrenches are made with 12 points so they can also be turned slightly and reset to remove bolts and nuts in close areas.

Note: Box wrenches completely enclose the nut or bolt head. They are also made in combination styles with a box end on one end and an open end on the other.

Socket wrenches are similar to box wrenches in that they also completely surround the bolt or nut, Figure 12-18A. Sockets are made in sets, detachable from solid or ratchet handles. A drive handle is shown in Figure 12-18B. *Hex socket head wrenches*, Figure 12-19, are made to be used with hex socket head cap and set screws. Figure 12-20 shows the Bondhus hex tool. This tool is

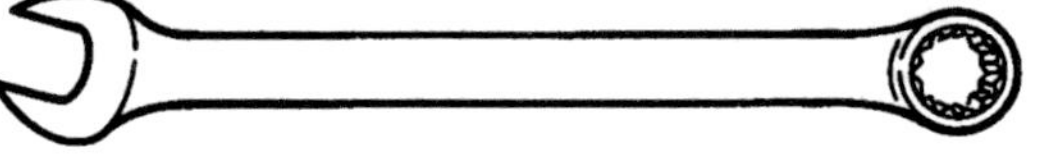

Fig. 12-16 Open-end wrench

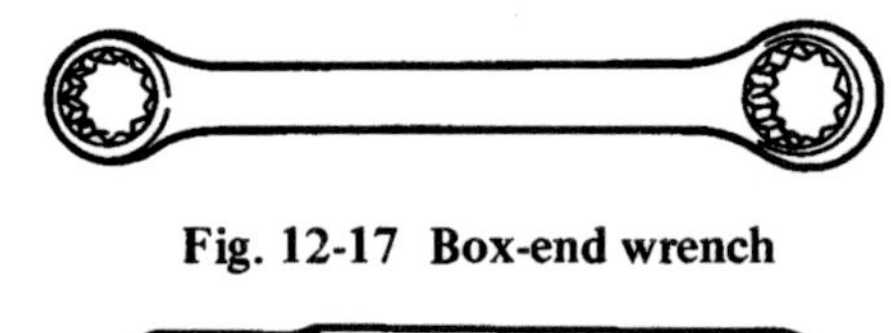

Fig. 12-17 Box-end wrench

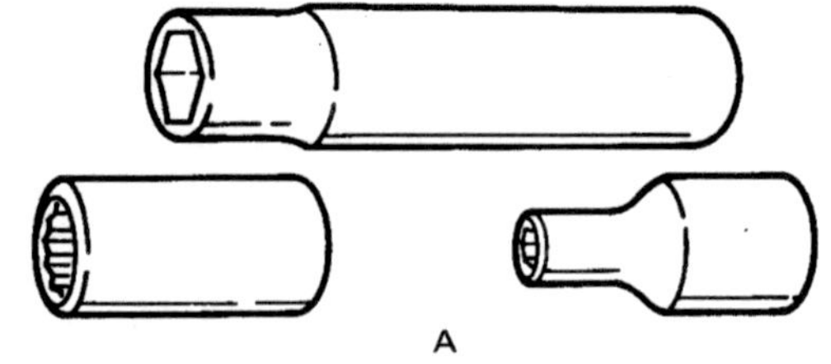

Fig. 12-18 (A) Various socket wrenches, and (B) a socket-wrench drive handle

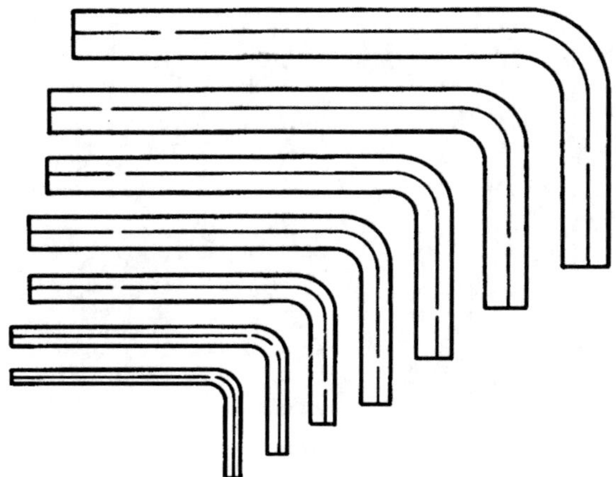

Fig. 12-19 Hex socket head wrenches

Fig. 12-20 Bondhus-style hex key wrench being used
(Bondhus Tool Company, Inc.)

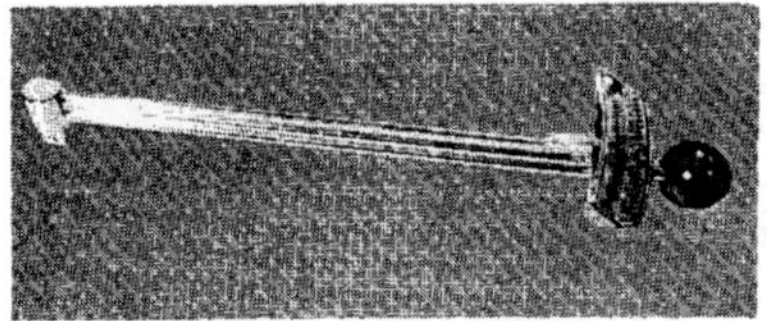

Fig. 12-21 Torque wrench

Fig. 12-22 Use of the torque wrench to correctly
tighten a bolt

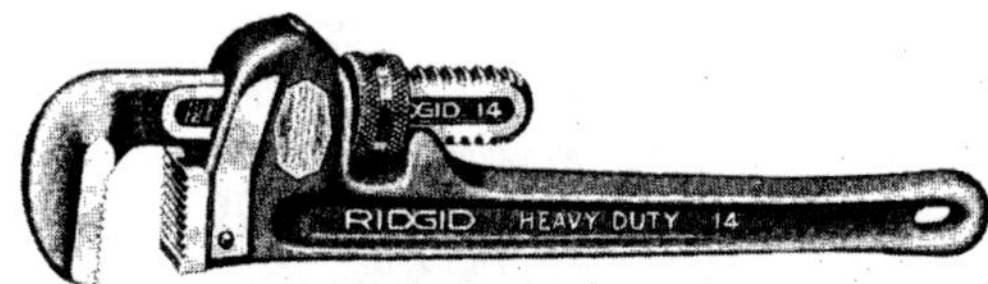

Fig. 12-23 Pipe wrench (The Ridge Tool Company)

Fig. 12-24 Strap wrench

designed to increase the speed of socket head
screw adjusting. These wrenches also come in
sets.

Ratchet wrenches are used to speed the
job of installing or removing fasteners. *Torque
wrenches* are designed to permit the application of a desired rotating or twisting power to
a fastener, Figure 12-21. This wrench has a
specially designed handle to read the torque
(rotation) applied. Sockets are usually used
with this special handle. Figure 12-22 shows
use of the tool.

Pipe wrenches are made in various
lengths and styles. They are used to assemble
or disassemble pipe and pipe fittings, Figure
12-23. Most have fixed teeth on the hook jaw
and replaceable teeth on the heel jaw. Pipe
wrenches range in size to fit 3/8-inch to 8-inch
pipe. Pipe wrenches should never be used as a
hammer or to lift or bend an object. They are
designed only to hold or rotate parts.

The *chain wrench* is designed for the same
purposes as the standard jaw-type pipe wrench.
The chain wrench may be used to advantage
in close places. *Strap wrenches* are used on
polished or specialty finished pipe or tube
parts to avoid teeth marks, Figure 12-24. The

Fig. 12-25 Spanner wrenches

wrench grips with a canvas strap to turn round objects.

The two types of *spanner wrenches* are fixed and adjustable, Figure 12-25. They are used on adjusting collars, lock nuts, rings, and spindle bearings. Figure 12-26 shows the pin spanner wrench used to tighten a part. The pins of the wrench are inserted into the part to be tightened or loosened.

Other spanner wrench styles are the hook and adjustable hook. The *hook-style spanner wrench* provides a hook or jaw to insert into slots of the part.

Use and Care of Wrenches

- Select the correct size wrench for the job.

- Never use an extension on a wrench to provide extra leverage.

- Always pull on a wrench handle and stand so you may prevent a fall if something slips or breaks.

- Always use a box-wrench style with a straight handle, if possible.

- Make sure the open-end wrench fits the nut or bolt so the wrench does not slip and round off the corners of the fastener.

- Stand to one side when pulling down on wrenches above your head.

- Never use a wrench in a way that if it slips, it could strike the hands or face.

- Use adjustable wrenches so the force is on the fixed-jaw closing and not spreading its action.

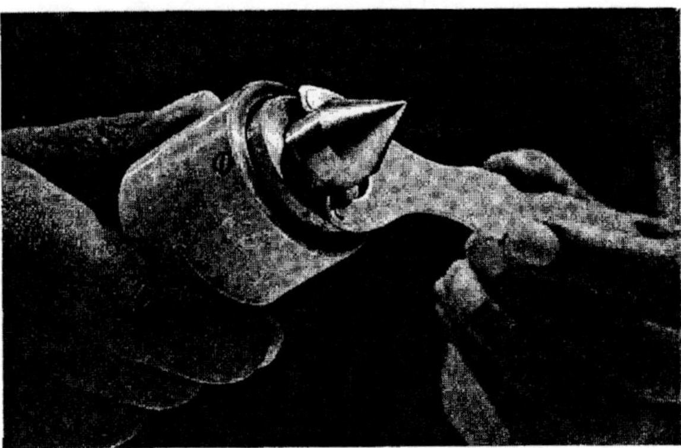

Fig. 12-26 Pin spanner wrench used to tighten a part

- Fully seat the jaw opening of the wrench on the nut or head of the bolt.

- A sharp blow with a soft-faced hammer on the wrench handle will possibly free a frozen nut or bolt.

SCREWDRIVERS

Screwdrivers are made in various sizes and lengths for very small precision work or large, heavy work, Figure 12-27. They are used mostly in standard and Phillips styles. Screwdrivers consist of a handle, usually plastic or wood, and a blade with the tip shaped to fit various screw slots.

The *standard screwdriver* is used to install or remove slotted-head screws or bolts. The screwdriver is gripped firmly by the handle. It is guided into the screw slot with the other hand.

Various shapes of screwdriver blade points are shown in Figure 12-28. The *Phillips screwdriver* tip has a 30-degree angle point. *Reed and Prince* has a point of 45 degrees. Screwdrivers are also made in ratchet and right-angle styles.

The *nut driver* is another tool that resembles the screwdriver, Figure 12-29. This tool has a handle and square blade to attach sockets to reach into close spaces. This tool is also made with a flexible shank to reach difficult places.

Most larger screwdrivers are made with a square shank. This allows a wrench to be used

Fig. 12-27 Screwdriver (The Stanley Works)

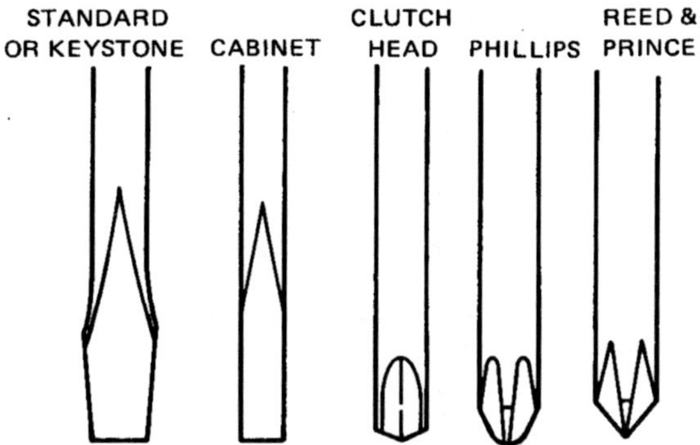

Fig. 12-28 Standard styles of screwdriver tips

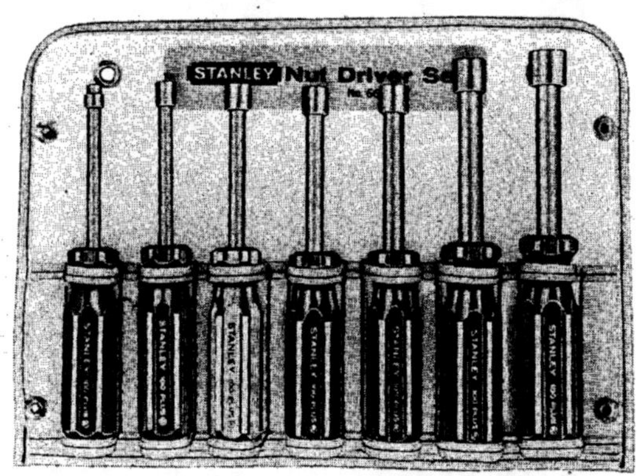

Fig. 12-29 Nut driver set (The Stanley Works)

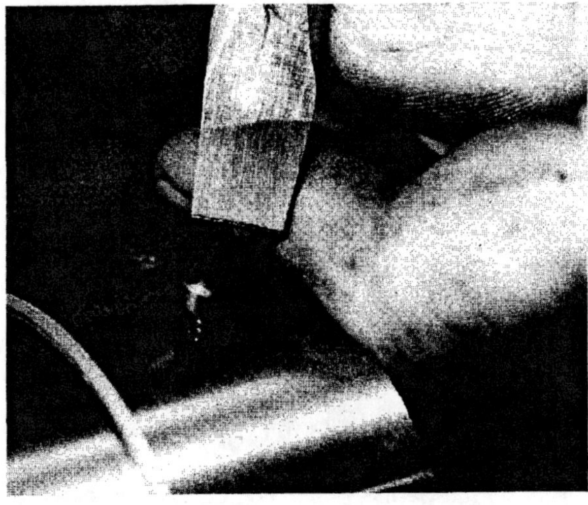

Fig. 12-30 Correct shape of screwdriver tip

for additional turning force while downward pressure is applied to the handle to keep the tip in the slot.

> **Caution:** When using a screwdriver, always hold the workpiece in the proper vise. This prevents possible personal injury if the workpiece slips and also provides for more accurate work.

Use and Care of Screwdrivers

- Use a screwdriver with the correct blade thickness for the screw slot. This prevents the blade from slipping out of the screw slot and damaging the screwhead. The width of the screwdriver tip should equal the length of the screw slot, Figure 12-30.

- Never use a screwdriver as a chisel, level, or wedge.

- Do not hammer on the end of the screwdriver handle.

BARS AND PUNCHES

The *pry bar*, Figure 12-31, is made of heat-treated alloy steel and is used as a lever to pry one part away from another. This bar is also called a pinch bar. *Pin punches* are used with a hammer to remove straight or taper pins in an assembly, Figure 12-32.

Fig. 12-31 Pry bar (The Stanley Works)

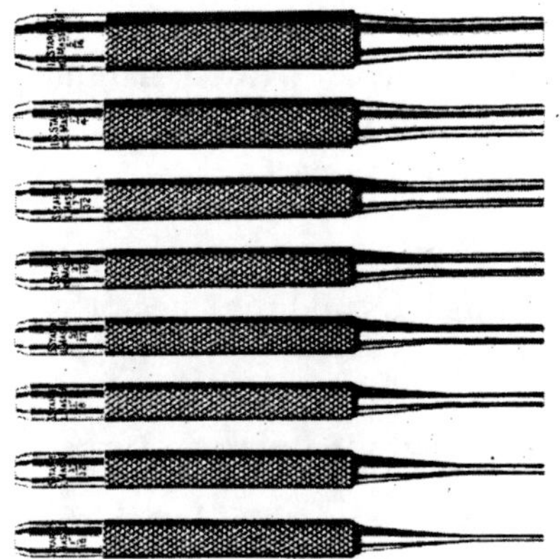

Fig. 12-32 Pin punch set (L. S. Starrett Company)

The *prick punch* is used for layout work to punch small indentations, Figure 12-33A.

The *center punch*, Figure 12-33B, is used after the layout is made to repunch the prick punch marks deeper. The deeper holes help start a drill bit into a workpiece.

The *aligning* (or drift) *punch* is used to line up holes in two workpieces, Figure 12-34. After hole alignment, a fastener is placed in the hole as the pin is removed.

Note: Punches may be resharpened as they become worn. The punch end must be kept cool while grinding so the cutting edge does not become soft. Punches are made of high-quality, heat-treated alloy steels to hold a very hard and tough cutting edge.

Transfer punches are used to transfer the location of a hole from a pattern to another workpiece, Figure 12-35. The punch has a center punch machined on one end. The punch is inserted through an existing hole. A hammer is struck on the end, and the punch produces the center of the hole in the workpiece.

Pullers are used to remove bearings and various hubs from shafts and housing, Figure 12-36. Figure 12-37 shows the internal puller being used. Always use the correct adapter to protect the shaft center hole, bridge a hole, thread into tapped holes, or assist in the installation of a part. Always pull against the pressed part of the workpiece.

Steel letters and numbers are used to mark and identify parts. These stamps are struck on the head with a hammer, Figure 12-38. The identation of the number or letter is made into the part.

> **Caution:** Wear safety glasses at all times when using bars and punches.

ARBOR PRESS AND SHOP PRESS

The *arbor press* is used with the hands to assemble one part to another, Figure 12-39. The part is placed on the press table and the

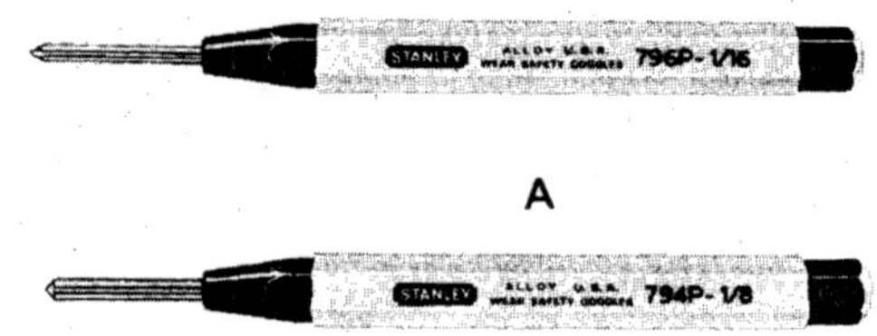

Fig. 12-33 (A) Prick punch, and (B) center punch (The Stanley Works)

Fig. 12-34 Aligning or drift punch (The Stanley Works)

Fig. 12-35 Transfer punches are used to transfer the center of a hole to another part.

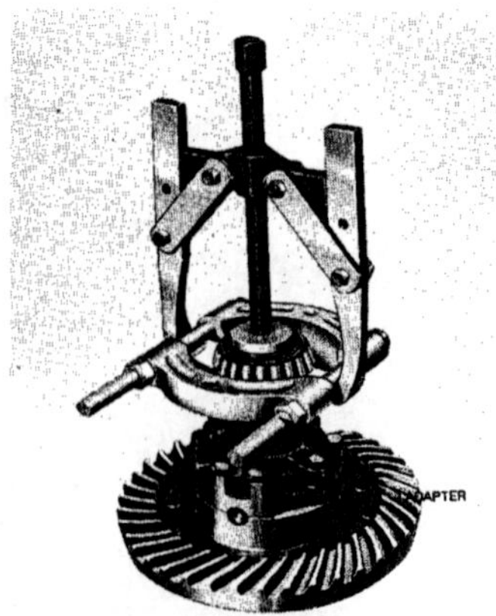

Fig. 12-36 External puller (OTC, Division of Owatonna Tool Company)

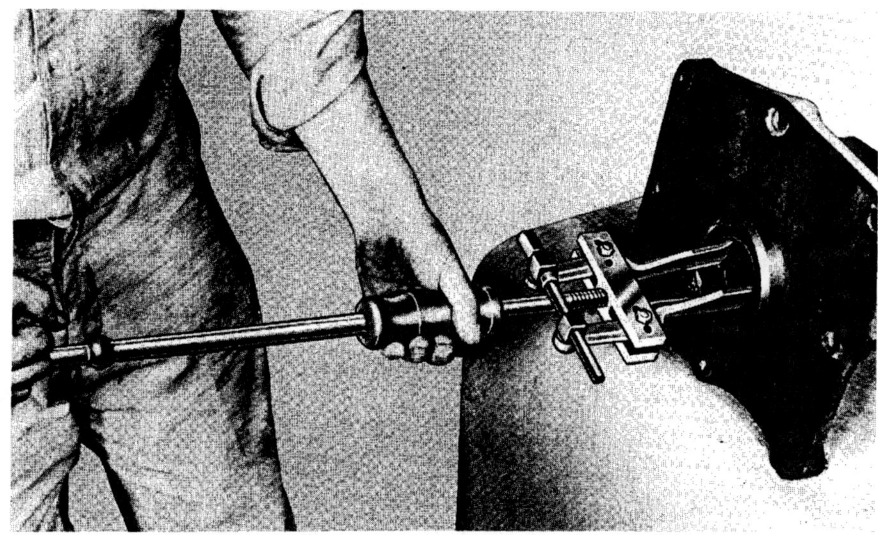

Fig. 12-37 Use of the internal puller
(OTC, Division of Owatonna Tool Company)

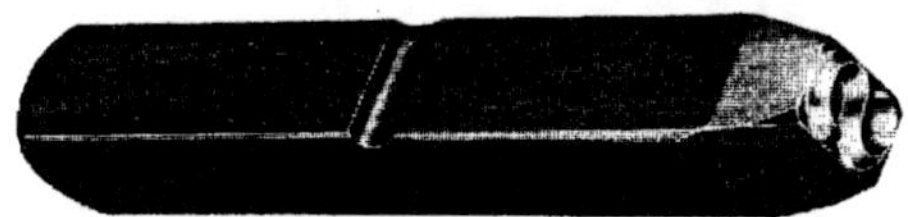

Fig. 12-38 Steel letters

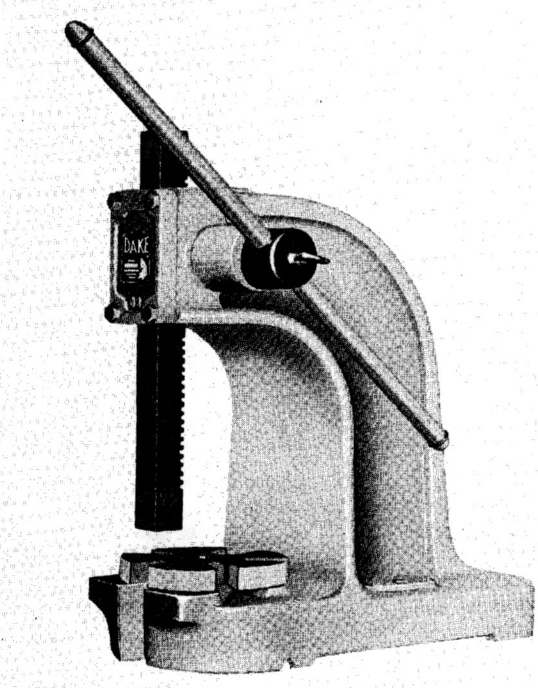

Fig. 12-39 Small, bench-type arbor press
(Dake Corporation)

ram is fed down with applied hand pressure to install or remove one part from another.

Figure 12-40 shows a hand-screw-operated shop press. Hydraulic-powered shop presses are also available. *Shop presses* are used to bend or form parts, straighten bent shafts, keyway broaching, and assembly of parts.

There are many other noncutting hand tools used in benchwork. Consult with a tool supplier or catalog to become familiar with the numerous tools available to assist the benchworker.

At times, the benchworker is required to design and make special tools to do special operations. Select the correct tool for the job and always use it safely.

Fig. 12-40 Hand-screw-operated shop press
(Dake Corporation)

REVIEW QUESTIONS

A.   Multiple Choice

1. The main difference between a standard and a Phillips head screwdriver is:

   a. the size of the shank.   c. the shape of the tip.
   b. the length of the shank.   d. the shape of the handle.

2. When is an adjustable, open-end wrench used?

   a. When the proper size open-end or box wrench is not available.
   b. When most of the bolt heads are metric.
   c. At all times in general-purpose work.
   d. Only when the bolt heads are large.

3. Which of the following tools is best used to install or remove smoothly finished chrome pipe?

   a. Pipe wrench   c. Chain wrench
   b. Strap wrench   d. Vise grips

4. Which of the following hammers is used for heavy work?

   a. Ball peen hammer   c. Sledge or mall
   b. Cross peen hammer   d. Nail hammer

5. The __________ punch is used to line up holes in two workpieces.

   a. pin   c. center
   b. prick   d. aligning

6. __________ are used to remove bearing and various hubs from shafts and housings.

   a. Pullers   c. Pipe wrenches
   b. Transfer punches   d. Spanner wrenches

B.   Short Answer

7. What is the recommended lubricant to use on the screw and nut assembly of a clamp?

8. Name the type of clamp that has two screw members.

9. What pliers are used for looping wire?

10. The main function of a __________ is to hold and turn nuts, bolts, cap screws, etc.

11. How are hammers sized?

ACTIVITY

1. From the following noncutting hand tools, demonstrate the proper care and use of each. Include safety precautions concerning the tool and the worker.

   a. Vise   g. Screwdriver
   b. Pliers   h. Pullers
   c. Hammer   i. Arbor press
   d. C clamp   j. Punches
   e. Open-end wrench   k. Adjustable wrench
   f. Spanner wrench

UNIT 13 CHISELS, FILES, SCRAPERS, AND ABRASIVE CLOTH

OBJECTIVES

After completing this unit, the student will be able to

- identify the commonly used cutting hand tools and abrasive cloths used in benchwork.

- describe the correct selection, use, and care of hand cutting tools and abrasive cloths.

INTRODUCTION

A certain amount of finishing and removal of material from machined surfaces is necessary to complete a part. This final fitting of the workpiece is done at the bench with hand cutting tools.

CHISELS

Chisels are made from heat-treated steels in various shapes. Chisel sizes range from 3/16 inch to 1 inch in flat blade width and in lengths from 5 inches to 12 inches. The chisels commonly used at the bench are known as cold, cape, round, and diamond point, Figure 13-1. They are considered the simplest of the cutting tools using a wedging action to cut and chip cold metal. The cutting edges are hardened. The head is tough and chamfered to prevent it from spreading or splintering from hammer blows. The cutting edge is placed on the material to be cut and the head is struck with a hammer.

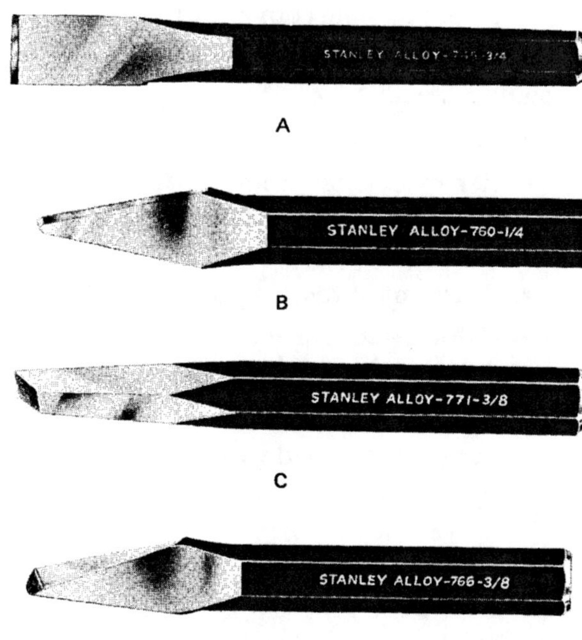

Fig. 13-1 Various chisel shapes: (A) Flat cold, (B) diamont point, (C) round nose, and (D) cape (The Stanley Works)

The *cold chisel*, also called a flat chisel, is used to cut metals and other materials. It is also used to shear off old, rusted bolt heads or nuts. Additional uses are:

- Cutting a design out of heavy sheet metal
- Cutting off small rods and bars
- Shearing heavy sheet metal in a vise
- Cleaning up rough castings

The *cape chisel* is used to:

- clean up battered keyways.
- cut some special keyways, channels, and slots in metal.
- divide work so that a flat chisel can finish, Figure 13-2.

The *round-nose chisel* is similar in shape to the cape chisel with one edge ground flat leaving the other edge a round cutting edge. Some uses of this chisel are to:

- cut round, half-circle grooves in metal, Figure 13-3.
- clean up round corners on rough castings.

The *diamond-point chisel* is used for the following:

- Cut V grooves
- Remove metal from a partially drilled hole to draw the drill back to the correct center position, Figure 13-4.
- Rotate out broken studs
- Cut holes in flat stock
- Clean out and finish square corners of a workpiece
- Cut out unwanted weld material
- Remove metal from a joint or crack to be welded

Care of Chisels

1. Chisels must be kept sharp. They are ground to an included angle of 60 degrees to 70 degrees, Figure 13-5. This angle can be less for softer materials.

2. Flat chisels give better shearing action to the cut if the cutting edge is convex

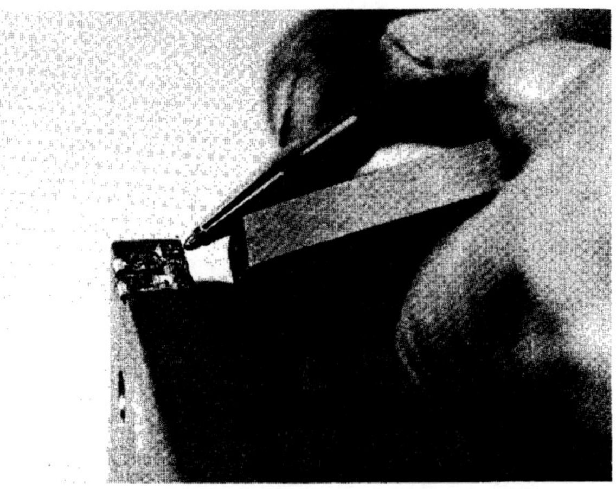

Fig. 13-2 Work is divided, or grooved, by the cape chisel and ready to be finished with a cold or flat chisel.

Fig. 13-3 Round-nose chisel used to cut a groove

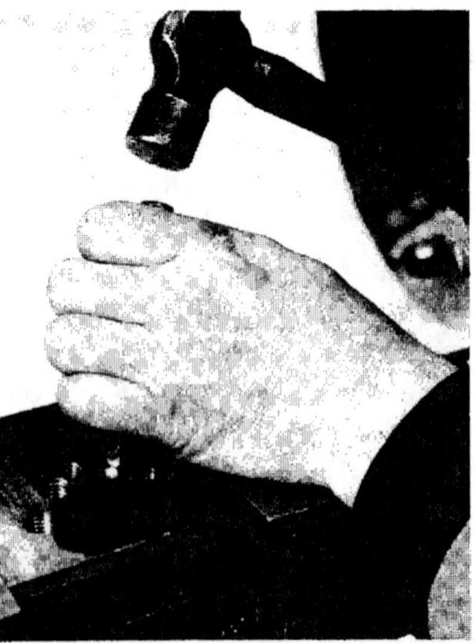

Fig. 13-4 Removing metal from a partially drilled hole to draw drill back to correct center position

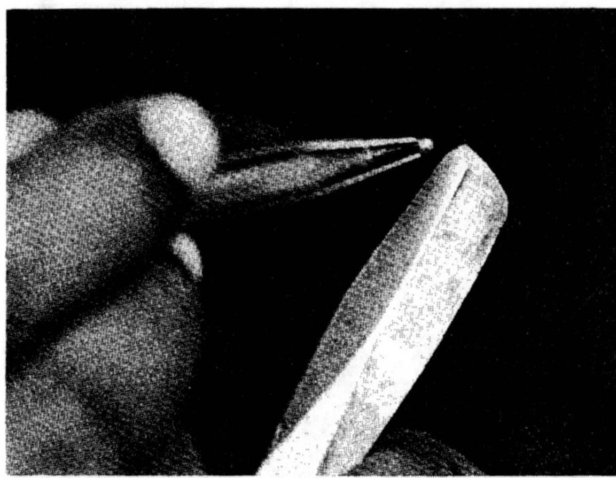

Fig. 13-5 Chisel point is ground to an included angle of 60 degrees to 70 degrees

Fig. 13-6 Dipping the chisel in cold water to keep the cutting edge cool

Fig. 13-7 Grinding the chisel

ground. With this edge, there is less chance for the corners to dig into the workpiece. Also, the pressure put on the tool is taken up through the stronger center part of the cutting edge.

3. When grinding a chisel, keep the cutting edge cool by frequently dipping it in cold water, Figure 13-6. Grind both sides of the chisel, taking equal amounts from each side. Move the chisel back and forth across the wheel face with moderate pressure, Figure 13-7.

4. Avoid too much grinding pressure so the cutting edge is not overheated. If it is overheated, the hardness is drawn out of the tool causing loss of cutting-edge hardness.

5. While in use, the head of the chisel frequently becomes mushroomed, Figure 13-8B. File or grind the chisel head back to its original shape, Figure 13-8A.

Using the Chisel

Chisels are used where machining is either difficult or impractical. Chisel cutting should normally be followed by a filing operation.

> **Caution:** Always wear goggles when using or grinding chisels. Protect other workers by placing a protective screen in back of work while chipping.

1. Fasten the laid out workpiece securely to the bench or in a suitable vise. Use soft jaws in the vise if the work has finished surfaces.

Fig. 13-8 (A) Properly dressed end of a chisel, and (B) end of a chisel slightly mushroomed

2. Set a chipping guard in place.

3. Select the correct chisel for the job, properly sharpened with the head in good shape.

4. Use a 3/4-pound to 2-pound ball peen hammer.

5. Grasp the small chisel with one hand in a steady, relaxed grip. The little finger is used to guide the chisel. The thumb and forefinger should be relaxed to take up the shock of the hammer blow, figure 13-9.

6. Swing the hammer with long swings and strike the chisel with firm, sharp blows, not hard blows. Watch the cutting edge of the chisel, not the head or anvil end. This makes directing the cut easier. When using larger chisels, all fingers should encircle the chisel as shown in Figure 13-10.

7. Begin to chisel at the outer edges of the piece and work toward the middle. This prevents tearing and breaking of the workpiece edges and corners, Figure 13-11.

8. Reset the chisel after each blow.

9. Cuts of about 1/32 inch (0.8 mm) to 1/16 inch (1.5 mm) deep should be made. The depth of cut is determined by the angle of the chisel to the work. Raise or lower the chisel angle to obtain the proper depth of cut. The higher the angle of the chisel, the deeper the cut. If held too low, the chisel may slip while chipping.

10. Large surfaces to be chipped should first have parallel grooves cut with a cape chisel. Refer to Figure 13-11. The spaces between the grooves should be slightly narrower than the width of the cold chisel to be used to cut the remaining metal.

11. Small work may be sheared by holding it in a vise and using the vise jaw as a guideline. Thin sheet metal should be backed up with a wood or metal plate to prevent bending.

12. Rivets and bolt heads are cut by placing the chisel cutting edge against the head

Fig. 13-9 Correct manner to hold the small chisel

Fig. 13-10 For heavy-duty cutting work, grasp the chisel as shown.

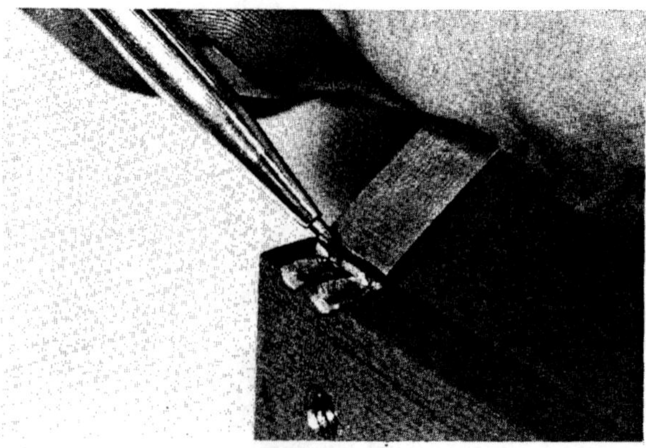

Fig. 13-11 Begin at outeredge and work toward the middle.

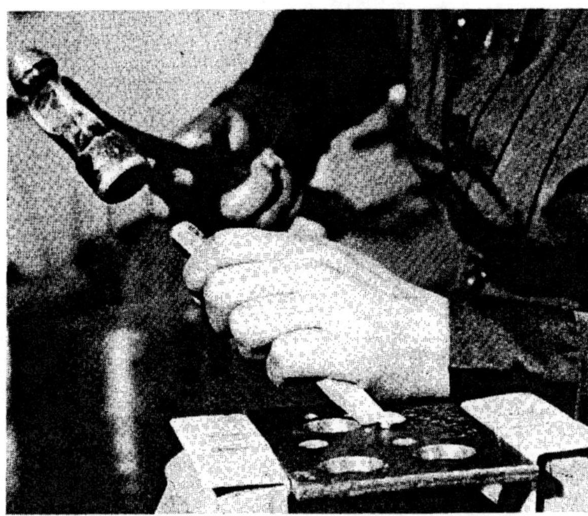

Fig. 13-12 Rivet being cut with a chisel

Fig. 13-13 File with a handle attached (Nicholson File Company)

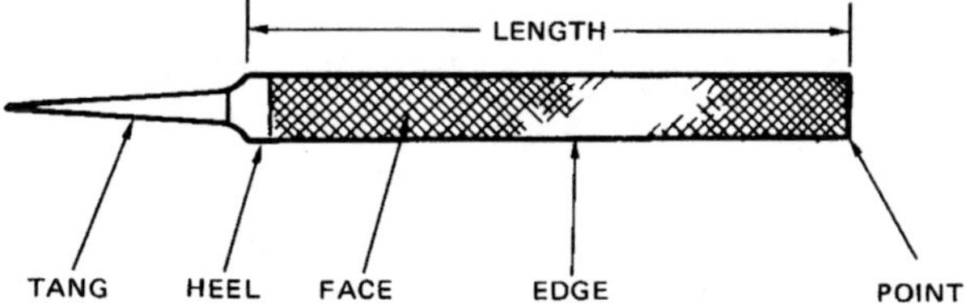

Fig. 13-14 Parts of the file

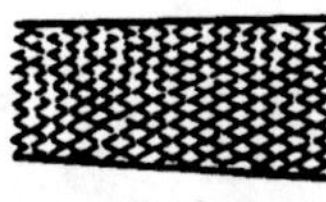

Fig. 13-15 File cuts

near the shaft, Figure 13-12. Strike the chisel sharply with the ball peen hammer.

13. Finish the chiseled work with a file or hand grinder.

> **Caution:** Always chip in a direction away from yourself and other workers. Put chisels back in their proper storage area when finished working.

FILES

A *file* is a piece of hardened, high-grade steel with slanting rows of teeth chiseled into its surface, Figure 13-13. These teeth cut and remove small amounts of material from a workpiece. Filing is most often used to smooth surfaces by removing sharp edges and burrs from machined parts. Files will not cut hardened steel or other materials harder than the file. The parts of a file are shown in Figure 13-14.

Files are not flat but made convex in their flat length. The shape allows cutting clearance for each tooth and allows a flat surface to be filed. If all the file teeth were in contact, too much downward pressure would be required to make the file cut. Also, too much forward pressure would be needed to make it cut.

Cuts of Files

The cut of a file must be considered when selecting a file for roughing, finishing, or draw-filing operations. The *cut of a file* is the coarseness of the teeth, the spacing of the teeth or the number of teeth per inch. Tooth spacing varies with the shape and length of the file. The spacing increases as the length of the file increases. The coarseness of a file can only be compared with files of the same length and shape.

The character of the file teeth is the cut. The four standard file cuts are shown in Figure 13-15.

- *Single cut* has a single set of diagonal rows of teeth.

CROSS-SECTION	NAME	SHAPE	CHARACTER OF TEETH	TAPER	GENERAL USES
	Flat	Rectangular	Usually bastard. Also second cut and smooth	Tapered in width.	A general-purpose file.
	Hand	Rectangular	One edge safe. Bastard, second cut and smooth.	Uniform in width.	Finishing flat surfaces.
	Pillar	Almost Square	One edge safe. Bastard, second cut and smooth.	Uniform in width.	Keyways, slots, narrow work.
	Warding	Thin	Usually bastard. Also second cut and smooth.	Width sharply tapered thickness uniform.	Filing ward notches in keys. Narrow work.
	Square	Square	Bastard, second cut and smooth.	Tapered	Enlarging holes or recesses. Mortises, keyways and splines.
	Three-Square	Triangular	Sharp edges. Bastard, second cut and smooth.	Tapered	Filing acute angles, corners, grooves, notches.
	Round	Circular	Usually bastard. Also second cut and smooth.	Tapered	Enlarging holes; shaping curved surfaces.
	Half-Round	Third-Circular	Usually bastard. Also second cut and smooth.	Usually tapered, sometimes blunt.	Concave corners; crevices, round holes.
	Knife	Knife-Shaped	Usually bastard. Also second cut and smooth.	Tapered curving to a narrow point.	Cleaning out acute angles, corners, slots.
	Aluminum	Flat Rectangular	Made in one cut only. Fast cutting teeth.	Tapered	Filing aluminum alloys and other soft metals.
	Aluminum	Half-Round	Made in one cut only. Fast cutting teeth.	Slightly tapered.	Filing aluminum alloys and other soft metals.
	Long Angle Lathe	Flat Rectangular	Made in one cut only. Both edges safe.	Slightly tapered.	Lathe work where smooth finish is desired. Also soft metals.

Fig. 13-16 File names, shapes, and uses
(Simonds Cutting Tools Division)

- *Double cut* has two sets of diagonal rows of file teeth. The first set of teeth is called the *over cut*. The other row of teeth, cut at a different angle, is known as the *upper cut*. The *upper cut* is finer than the *over cut*.

- *Rasp cut* is a pattern by which each tooth is individually formed. The teeth are separate and disconnected.

- *Curved cut* teeth curve across the face of the file.

File Shapes

Files are generally known by their cross section, shape, or use. They are made in hundreds of shapes and sizes for many job applications. A flat file is used on flat work. The mill file is a general, all-purpose file. Half-round and round files are used on curved surfaces. The most commonly used file shapes are shown in Figure 13-16.

A *mill file* is used for smoothing lathe work, drawfiling, or fine precision work. Mill files are always single cut. One edge is without teeth, known as the *safe edge*. *Flat files* are used for general-purpose work; double cut for rough work and single cut for smooth finishing. *Square files* are used to enlarge rectangular-shaped holes and slots. Round files enlarge round-shaped holes.

Half-round files are used for a wide range of different jobs.

Swiss pattern files, Figure 13-17, are used to fit parts of a delicate assembly.

Three square (triangle) files are used to file corners square and filing burrs from 60-degree thread grooves. They are also used to sharpen chain saw teeth.

Filing Operations

The following are descriptions of basic, bench filing operations.

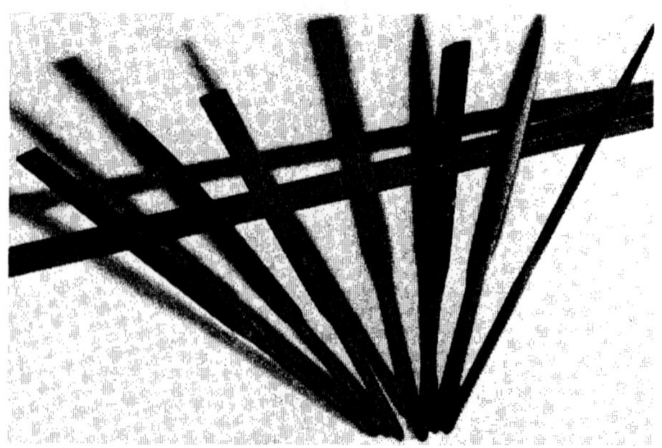

Fig. 13-17 Swiss pattern files

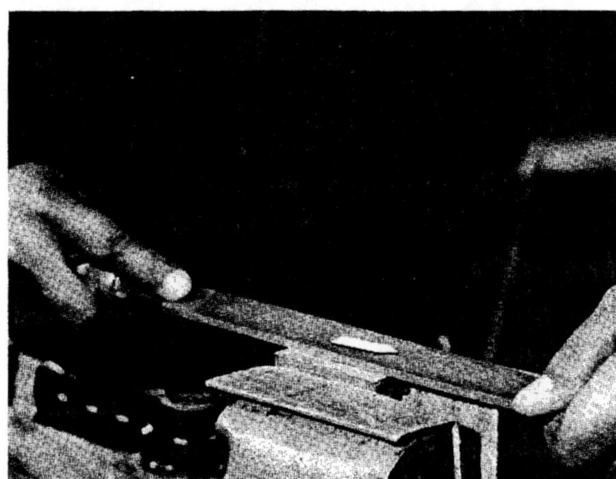

Fig. 13-18 Flat filing (Note: Arrow shows direction of filing strokes)

Fig. 13-19 Drawfiling

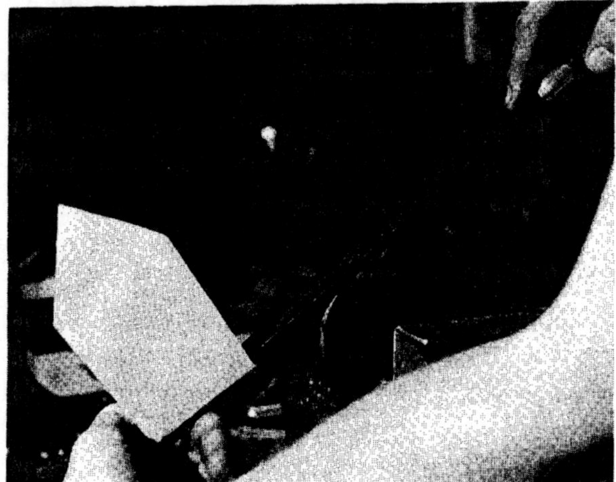

Fig. 13-20 When round-corner filing, start in the position shown and finish in the direction of the arrow.

Straightforward, flat filing is a filing operation where the file is pushed straight across the work, Figure 13-18.

Hold the thumb and forefinger stretched apart and pressed evenly on the file. This provides a more even pressure throughout the length of the file. The file tends to remain flat. The greatest fault in filing a flat surface is allowing the file to rock or see-saw. This produces a curved surface rather than a flat surface. To avoid this, keep the body still and pivot the arms about the shoulders.

Also, try to remove less material in one stroke. Be sure to use a clean, sharp file. For very accurate filing, the file tip should be held by just the thumb and index finger. This grip allows for maximum control.

Drawfiling is a filing operation to produce an even-grained, smooth, true flat surface. This is done by pulling or drawing the file at right angles across the work, Figure 13-19.

In drawfiling, the file is held with the fingers and thumbs on the edge of the file, as shown in Figure 13-19. The file is alternately pulled towards the body with a pressure stroke, then pushed away across the work with relaxed pressure. One advantage of drawfiling is that the file can be held steadily.

Usually, a single-cut, mill bastard file or a long-angle lathe file should be used for drawfiling. This type of file avoids scoring to the work because of a truer shearing or shaving action to the cut. Make sure not to apply more pressure in the middle of the cut causing a hollow in the workpiece. Chalk may be used to keep the file clean and produce a better finish.

Round-corner filing, Figure 13-20, is a filing operation where the file is pushed down and around the corner of the curved work. The point is then lifted and the handle pushed downward throughout the cutting stroke.

Filing Procedure

The work must be held firmly and rigidly to prevent vibration and chatter. The top of the vise should be on the same level as the elbow when the arm is bent. Protect finished work surfaces held in the vise with pieces of copper, brass, zinc, wood, lead, or leather.

1. Place your feet about 24 inches apart. A full, free swinging movement of the arm and shoulder should be used. Avoid separate movement of the wrist and elbow.

2. Grip the file handle with one hand and the tip with the other. Rest the file handle in the palm of the hand.

3. For heavy stock removal, grip the file tip with the hand and the ball of the thumb, pressing on the top of the file.

> **Caution:** Never use a file without a tight-fitting handle. The tang is sharp and can cause severe injury.

4. The teeth of the file only cut on the forward stroke. Pressure should not be applied on the return stroke.

5. The file should be moved on forward strokes in an almost straight line. The pressure is first applied to the tip end, then both hands through the middle of the stroke.

6. Pressure is applied at the end of the stroke near the heel of the file.

7. With practice and patience, it is possible to file a surface true and square. Check the surface with a straightedge or precision square.

Note: Remember, very accurate filing should only be done if a machining operation is not available or is inconvenient.

Cleaning the File

File teeth can become clogged with chips between them. This causes the file to slide over the work instead of cutting. This condition is known as *pinning*. The *pins* or chips in the file teeth may be removed with a file brush or file card, Figure 13-21. Rubbing chalk between the teeth prevents pinning and produces smoother finishes. Also, a brass or copper bar pushed parallel across the teeth from edge to edge removes the pins. An impression of the file teeth is cut into the bar, thus removing the pins.

Fig. 13-21 File brush (Nicholson File Company)

Care of Files

- Wrap files separately in protective paper when in storage.

- Place bench files not in immediate use in suitable racks to protect their sharp cutting edges.

- Break in a new file by gently using it on flat surfaces of soft metals such as brass, bronze, or smooth cast iron. A new file should not be used on narrow, sharp corners such as sheet metal.

- Keep files free from rust and avoid getting them oily. A slippery file can cause serious injury. Oil makes the file slide across the work without any cutting action. Even the oil from the fingers can cause filing problems, especially on finished cast iron.

- Files can be resharpened with a sulfuric acid solution of three parts water to two parts acid. Always add the acid to the water. Place the file in a clean container. Fill the container with the solution to the top of the file teeth. Remove the file in 1 to 2 hours.

- Files must not be used as pry bars. The file metal is very brittle and may break causing serious injury.

File Selection

Each file made is designed for a certain kind of work. The size and type of work must be considered when selecting the correct file shape, length, cut and coarseness.

The *double-cut hand file* cuts metal quickly for rough work. A 10-inch or 12-inch *bastard file* is used for rough filing at the bench.

A *second-cut file* is used to bring the work surface closer to size. The *single-cut mill file* is used for finish cuts and smoothing the edges of sheet metal. The smooth and dead smooth files are used to produce finer finished surfaces.

SCRAPERS

Scraping is the process of removing thin shavings of metal from a workpiece surface to make it flat and smooth. Hand scraping is slow, expensive, and seldom done today. Power scrapers are used to speed up the scraping operation.

Scraping removes the high spots left after machining to accurately fit one part to another. These high spots are caused by the vibration between the cutting tool and the workpiece during the cutting operation. Precision grinding is a better method to produce true, flat surfaces on a workpiece. Scraping is also done to provide small oil pockets between mating surfaces.

Scraping Tools

Scraping is done with tools called *scrapers* by either pushing or pulling, Figure 13-22. The *flat* and *hook scrapers* are used on flat surfaces. The *three-cornered* and *half-round scrapers* are used on curved surfaces.

The cutting end of the scraper should be round slightly, Figure 13-23. This provides better control of the cutting action and reduces the chance of the corners digging into the work. The scraper handle is held under the arm and steadied with both hands. Pressure is applied to the cutting edge from the body and the arm. The cutting strokes should be from 1/4 inch (6.3 mm) to 1/2 inch (12.7 mm) in length.

Spotting

A surface plate is used to find the high spots on the work surface. This process of finding the high spots is called *spotting*. The top of the surface plate is lightly coated with a very thin film of Prussian blue or Venetian red paint.

The surface to be scraped is laid on the surface plate and moved back and forth. High spots on the workpiece are marked by the paint from the surface plate. These high spots are then scraped off with the scraping procedure.

After scraping, the work surface is tested again on the surface plate. Repeat these operations until the marks on the workpiece increase and are evenly spread over the work surface. Turpentine may be used with close finish scraping to show the high spots which appear to be bright. The turpentine also acts as a cutting lubricant and gives better scraping results.

Scraping is also used as a decoration to remove undesirable finish marks on flat

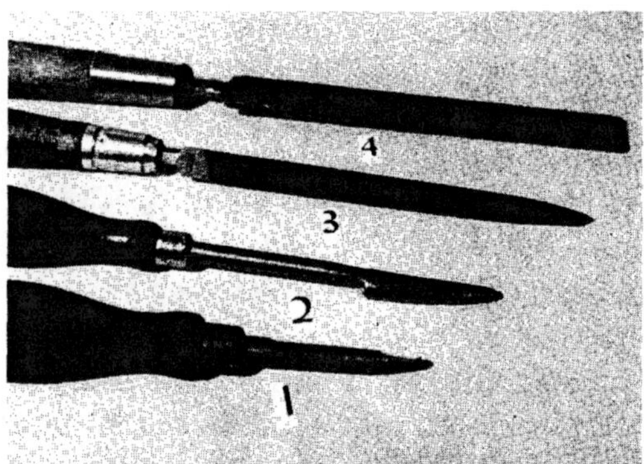

Fig. 13-22 Commonly used scrapers: (1) Hook, (2) curved, half-round, (3) three-cornered, and (4) flat

Fig. 13-23 Flat scraper with slightly rounded cutting edge

machined surfaces. *Frosting* or *flowering* are two types of patchwork or checkerboard scraping designs used to decorate. This process is done by scraping off high spots. Precision scraping is an art. It is best learned through experience by working with a master.

Deburring is done with various forms of scrapers, Figure 13-24. Sharp corners and edges are removed from the workpiece. The *countersinks* shown in Figure 13-25A and B are deburring tools for removing the sharp edges from drilled holes and preparing them for tapping.

ABRASIVE CLOTH

Polishing is done at the bench with the use of an *abrasive cloth*. Abrasive cloth produces a smooth, polished surface by removing very small amounts of material.

Abrasive cloths are coated with either natural or artificial abrasives. Emery is the most commonly used natural abrasive for abrasive cloth. Silicon carbide and aluminum oxide are artificial abrasives that are rapidly replacing emery. Abrasive cloth is available in either shop rolls or 9″ x 11″ sheets, the same as abrasive paper. Strips can be torn lengthwise from the sheets. The cloths are made in a number of grades according to grit size to remove varying amounts of material and to produce desired finishes.

Emery cloth is made by cementing the screen-sized grit particles to a cloth backing. Figure 13-26 shows other styles of abrasive polishing materials.

Grade Sizes and Applications

Different grades of emery cloth range from fine to coarse. The coarse grades are used for removing greater amounts of materials where the finish is not important. The coarse grains cut more rapidly. Medium grades are used to remove scratches and slight tool marks. Fine grades are used for polishing and finishing. Light, clean up work of tools, rust removal, and hand polishing of nonplated

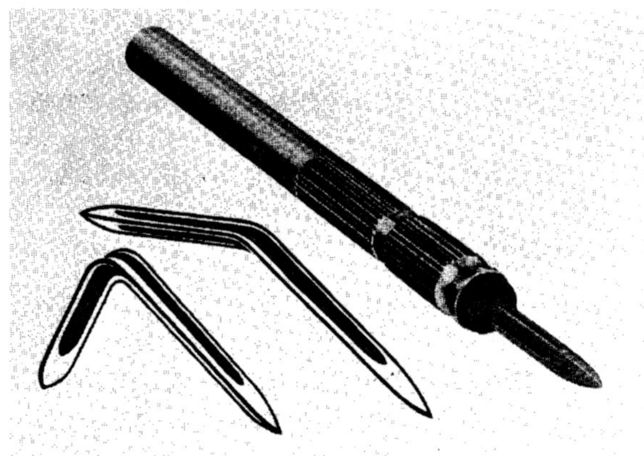

Fig. 13-24 Adjustable deburring scraper with a straight blade and two angle blades (Titan Tool Supply Company)

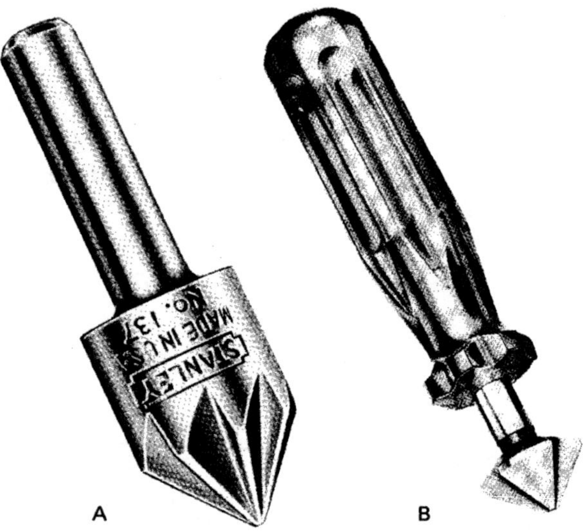

Fig. 13-25 (A) Machine countersink, and (B) hand countersink (A) The Stanley Works (B) Swiss Precision Instruments, Inc.)

Fig. 13-26 Various abrasive polishing materials (Behr Manning, Division of Norton Company)

	Grade	Mesh Size	Emery Number
	Fine	180	3/0
		150	2/0
		120	1/0
	Medium	100	1/2
		80	1
		60	1 1/2
	Coarse	50	2
		40	2 1/2
		36	3

Table 13-1 Comparison chart for grades of abrasive cloth

metal surfaces are other uses. Table 13-1 compares fine, medium, and coarse grades of abrasive cloth with mesh size and emery number. The larger the mesh-size number, the finer the emery cloth. The smaller the mesh-size number, the larger or coarser the abrasive grains.

How to Use Abrasive Cloths

Abrasive cloths are not intended to remove material to make fits. They are, however, used at times to remove small amounts of material and tool marks to make a close fit possible.

Caution: Abrasive-cloth edges are sharp and can cause severe cuts to the benchworker. Use abrasive cloth with care.

1. With the work held securely, place the emery cloth on the work and fold one end over the end of a file, Figure 13-27.
2. Move the emery cloth back and forth across the work using the file as a backing. Avoid a rocking motion on flat surfaces. A metal or hardwood block may be used rather than the file as a backing for the abrasive cloth. Greater control is obtained with the file to produce a flat surface.
3. Save worn emery cloth to use to produce finer finishes. Oil improves the polishing operation by reducing scratches. The work surface is also left in an oiled condition to resist rust longer.

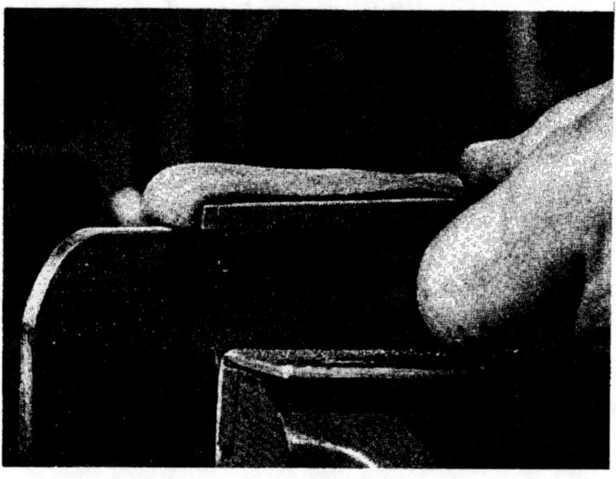

Fig. 13-27 **Proper way to place emery cloth on work**

REVIEW QUESTIONS

A.   Multiple Choice

1. The flat, cold chisel should have the cutting edge slightly ground in which of the following ways?
   - a. Convex
   - b. Concave
   - c. Square
   - d. Triangular

2. The part of a chisel to watch while chipping is the:
   - a. head or anvil.
   - b. body.
   - c. cutting edge.
   - d. hammerhead.

3. After continued use, the head of a chisel can become ____________ , causing a safety hazard.
   - a. flat
   - b. mushroomed
   - c. squared
   - d. tapered

4. The cutting edge of a cold chisel for general-purpose cutting should be at an included angle of:
   - a. 20 degrees to 30 degrees.
   - b. 30 degrees to 40 degrees.
   - c. 50 degrees to 60 degrees.
   - d. 60 degrees to 70 degrees.

5. A mill file is which of the following?
   - a. Square
   - b. Milled
   - c. Double cut
   - d. Single cut

6. Files should always be used with which of the following?
   - a. Speed
   - b. Oil
   - c. Pins
   - d. Handles

7. Files are best cleaned with a:
   - a. file cleaner.
   - b. brush card.
   - c. file card.
   - d. bench brush.

8. The process of scraping is used mainly to produce work surfaces that are:
   - a. flat and smooth.
   - b. square and clean.
   - c. even-grained.
   - d. spotted.

9. Polishing is done at the bench with the use of:
   - a. countersinks.
   - b. Prussian blue paint.
   - c. abrasive cloth.
   - d. turpentine.

ACTIVITIES

1. Demonstrate the correct procedure to sharpen and use a cold chisel.

2. Properly select a file and correctly use it to make a flat surface and remove sharp edges.

3. Demonstrate the hand scraping of a surface.

4. Demonstrate the proper use of abrasive cloth.

UNIT 14 DRILLS, REAMERS, BROACHES, AND SAWS

OBJECTIVES

After completing this unit, the student will be able to

- identify the commonly used hand cutting tools used in benchwork.

- describe the correct selection, use, and care of commonly used hand cutting tools.

INTRODUCTION

This unit discusses the hand cutting tools used at the bench that remove greater amounts of materials in finishing operations. Because of its design, the hand reamer is mentioned in this unit even though it only removes a small amount of material with every cut.

All cutting tools must be properly stored to protect the hard, sharp cutting edges from damage. Care must also be taken to keep tools from rusting.

DRILLS

Drilling is the process of making a circular hole into or through a piece of material, Figure 14-1. *Chamfering* removes the outer sharp edges of a hole and prepares the hole for tapping. A countersink is used in this operation, Figure 14-2. *Countersinking* is the term applied to making the end of a drilled hole to receive the 82-degree, taper-shaped head of a screw, Figure 14-3. *Counterboring* is the process of enlarging the upper part of

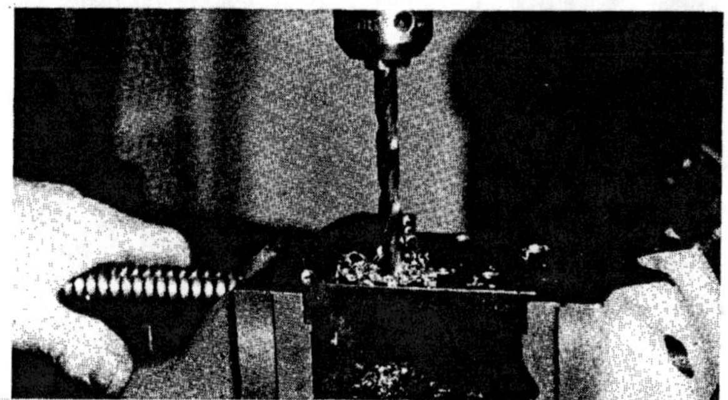

Fig. 14-1 Drilling a hole

Fig. 14-2 Countersink tool used to remove sharp corners from the edge of the drilled hole

Fig. 14-3 The drilled hole is countersunk to receive or fit the flathead screw.

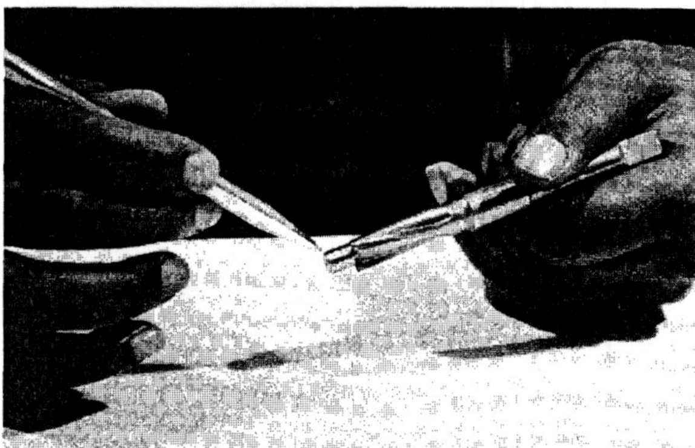

Fig. 14-4 The pilot of the counterbore is shown. The cutter above the pilot removes metal from the upper part of the drilled hole.

Fig. 14-5 Combination drill and countersink

Fig. 14-6 Twist drills

the drilled hole to receive the head of a bolt or cap screw, Figure 14-4. In the *spotfacing* process, the top surface of the hole is machined flat and square to the hole to receive a nut or bolt.

The *combined drill and countersink*, Figure 14-5, normally called a *center drill*, is used to spot drill holes before drilling. It is also used to drill 60-degree center holes in the end of work to be supported in the lathe.

Twist drills, Figure 14-6, are made in various styles, lengths, and in sizes from 0.0059 inch to 3 1/2 inches in diameter. The parts of the twist drill are shown in Figure 14-6A. They are available in standard sets such as:

- Fractional sizes in 1/64 inch — from 1/64 inch to 1/2 inch

- Letter sizes — A (0.234 inch) to Z (0.413 inch)

- Number sizes — #80 (0.0135 inch) to #1 (0.228 inch)

- Metric sizes — 0.35 mm (0.0138 inch) to 12.5 mm (0.4921 inch)

Twist drills are made of the following materials:

- High-carbon steel — for use in home shop

- High-speed steel — for general-purpose drilling

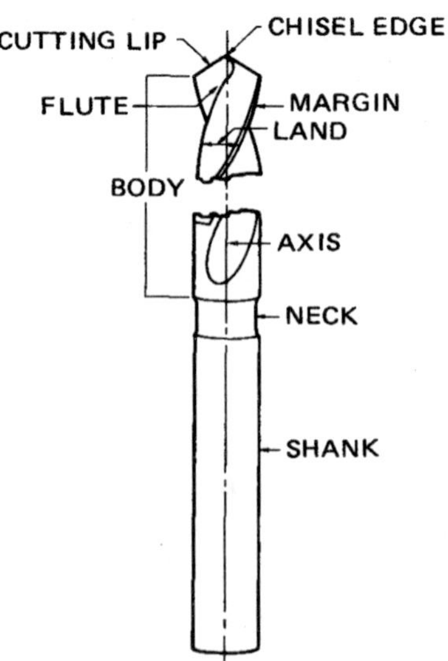

Fig. 14-6A Parts of a twist drill

- Cobalt high-speed steel — for drilling hard materials
- Carbide — for drilling harder, tougher materials than above

The benchworker uses drills with straight shanks. Usually these drills are held in an electric powered, hand drill motor.

Drilling speeds vary when drilling different kinds of materials. Also, the type of material from which the drill bit is made affects the drilling speed. Hard, tough materials must be drilled with slower speeds and heavier feeds. Softer materials are drilled with faster speeds.

Twist Drill Sharpening

For general-purpose drilling, the drill point should be ground as shown in Figure 14-7. As twist drills are used, the cutting edges become worn and dull.

Twist drills are best sharpened on a drill grinding machine. The machine grinding operation produces an accurate drill point quickly.

To regrind the twist drill point, observe the following important factors:

- Both cutting edges must be ground to the same length and angle.

- The cutting edges or lips must have correct clearance. This clearance is ground to lower that portion of the drill material behind the cutting edge.

- The location of the chisel edge at the center of the drill must be at the required included angle of 135 degrees, Figure 14-7A.

- Grinding a drill by hand requires practice.

Drill Grinding Procedure

1. Wear safety glasses and a face shield at all times.
2. With the wheel stopped, inspect the cutting surface of the grinding wheel. If necessary, remove dull abrasive grains and wedged grinding chips with a wheel

dresser. Dressing a wheel is further discussed in Unit 18.

3. Adjust the tool rest of the grinder to within 1/16 inch (1.5 mm) of the wheel face.
4. Rest the index finger of one hand on the tool rest and hold the drill near the point with the thumb and index finger, Figure 14-8. Hold the drill bit near the shank end with the other hand. The shank end

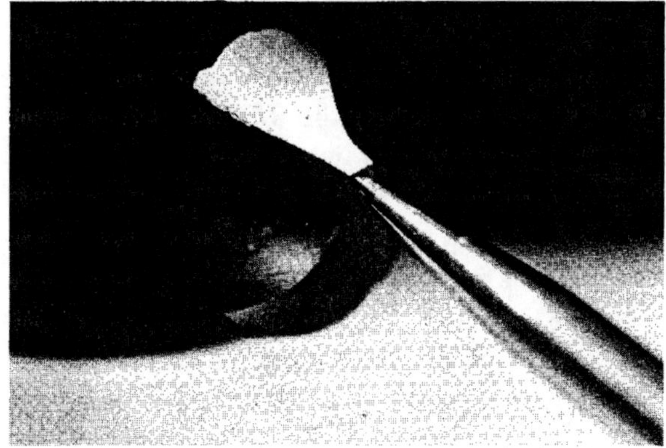

Fig. 14-7 Properly sharpened twist drill

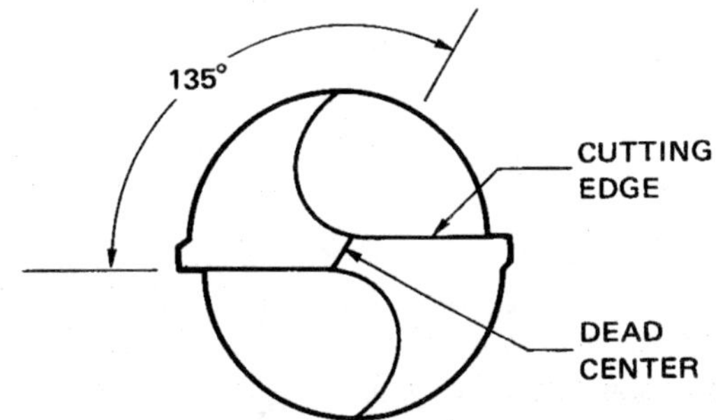

Fig. 14-7A

Fig. 14-8 Correct method of holding a twist drill to the grinding wheel for sharpening

should be level with or slightly lower than the point end, Figure 14-8A.

5. Move the shank of the drill around to an angle of approximately 59 degrees with the cutting face of the grinding wheel as in Figure 14-9.

6. Scribe a line on the tool rest at 59 degrees to the wheel face. This helps as a guide to hold the drill at the correct angle, figure 14-10.

7. Start the grinder.

8. Hold the drill so the cutting lip is parallel with the cutting face of the wheel.

9. Press the cutting lip of the drill against the rotating grinding wheel.

10. Slowly lower the shank end of the drill bit. At the same time, press the clearance area behind the cutting edge into the wheel. This inward pressure of the drill toward the wheel, combined with slowly lowering the shank, develops the correct clearance (8 degrees to 12 degrees) behind the cutting edge. Do not rotate the drill bit while lowering the shank to grind the clearance.

Note: Too much clearance weakens the cutting edges. Not enough clearance causes the drill to simply turn, rub, and dull quickly. Without clearance the cutting edge cannot cut or bite into the work. The correct amount of clearance is easily maintained by checking the chisel angle (125 degrees to 135 degrees) of the point.

11. Remove the drill from the grinding wheel. Rotate the drill bit to the other cutting lip. Grind this lip in the same manner as described in Steps 4 through 10.

12. Measure the lengths and angles of both cutting lips with a drill point gage, Figure 14-11. Both cutting lips must measure the same length and angle.

13. Repeat the grinding of each lip alternately until the drill point is free of wear. Both cutting lips must be sharp, the same length, and the same angle.

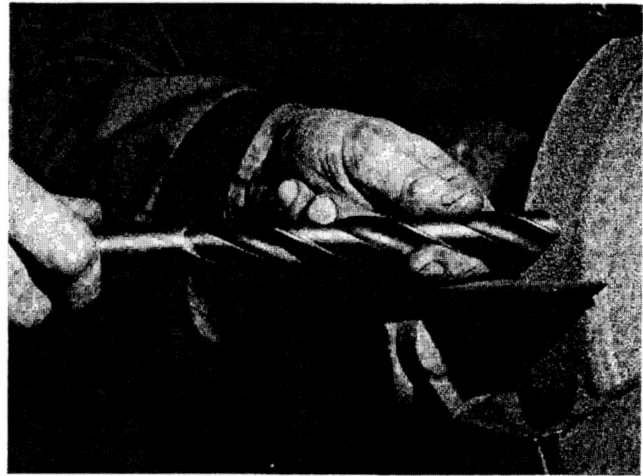

Fig. 14-8A

Fig. 14-9 Set the drill bit at 59 degrees with the face of the grinding wheel.

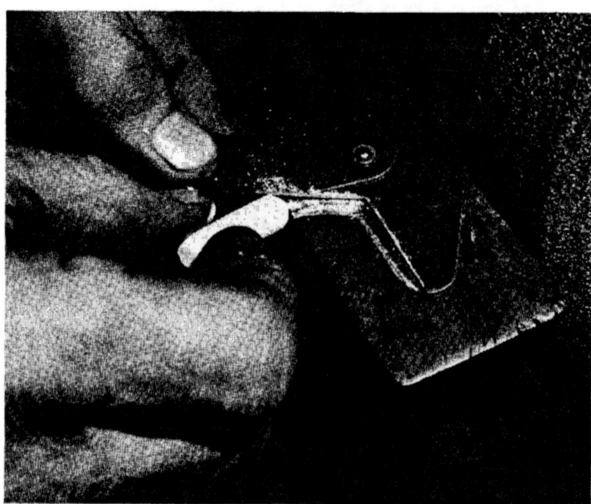

Fig. 14-10 Scribing a line on the tool rest as a guide

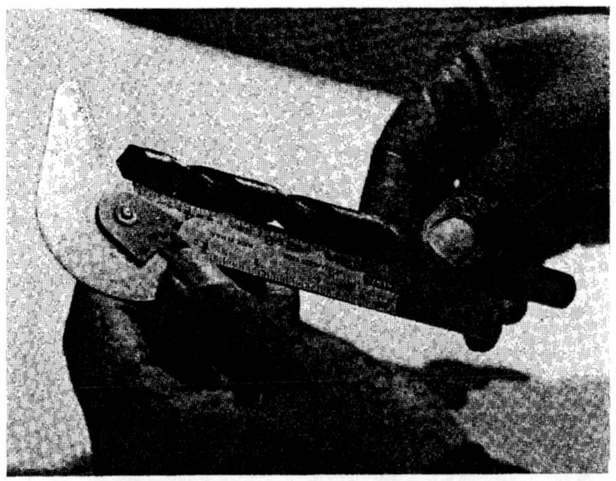

Fig. 14-11 Drill point gage used to
measure the lengths and angles of both cutting lips

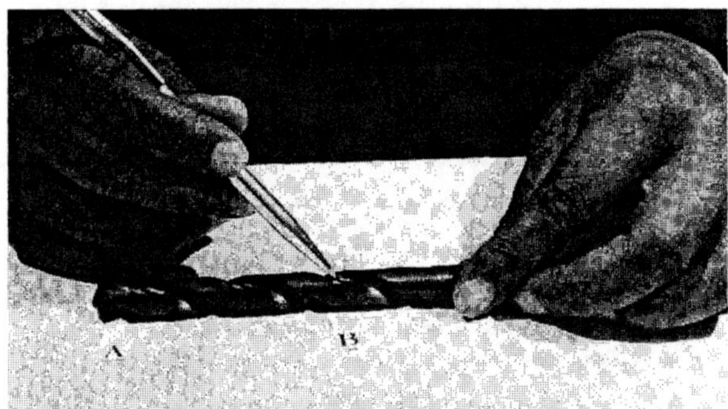

Fig. 14-12 Back taper is ground on the twist drill diameter. The drill is smaller at B than at A.

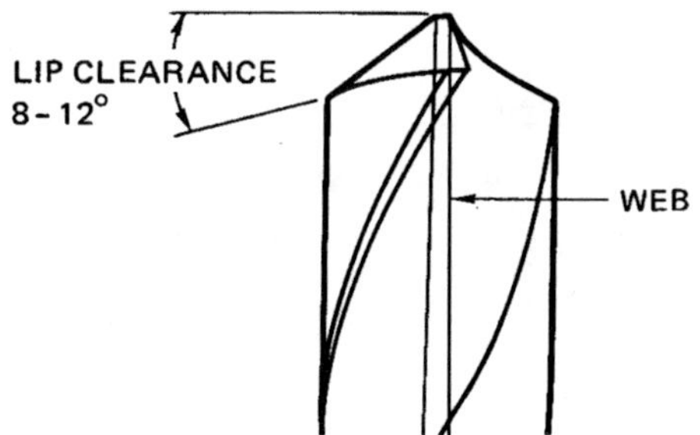

Fig. 14-13 The web of
the drill is thicker toward the shank.

Note: Keep the drill point cool to the touch
by frequent dipping in water.

All drills tend to drill oversize by some amount. This amount depends greatly on the grinding accuracy. Drill bits are made with a small amount of taper towards the shank. This taper provides clearance to the drill in a hole to reduce friction and heat, Figure 14-12.

Thinning the Web

As a drill bit becomes shorter from repeated regrinding, the web thickness increases towards the shank to make the drill bit stronger and more rigid, Figure 14-13. This thickness causes greater drilling pressures because of the longer chisel edge which has no actual cutting action. The web must be thinned evenly with the same amount of material removed from each cutting edge. The web can be thinned and the chisel edge shortened by grinding as shown in Figure 14-14.

Drill Point Shapes for Other Materials

Drills used to drill brass must have the cutting edges dubbed or ground parallel with the axis of the drill, Figure 14-15. There can be no hook at the cutting edge to cause the drill bit to dig or be pulled into the brass material. Soft materials are best drilled with a small 90° point angle. Hard, tough materials are best drilled with a flat 135° point angle,

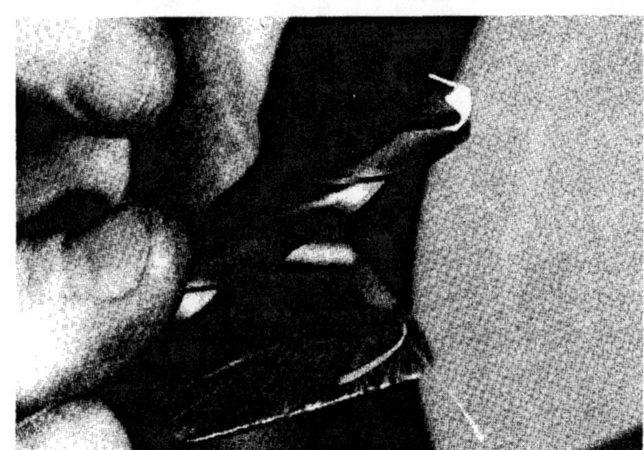

Fig. 14-14 Web is thinned on one side. The other side must be thinned the same amount on the grinder.

Fig. 14-15 Drills ground with the cutting edge parallel with the drill axis are used to drill brass.

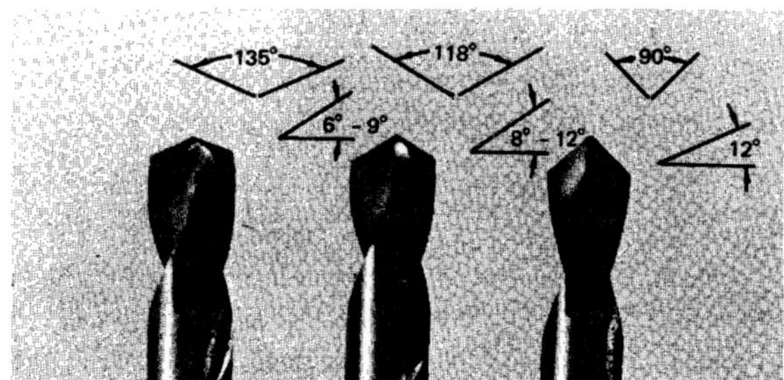

Fig. 14-15A

Figure 14-15A. Drill bit sizes can be checked quickly and easily with the use of drill gages, Figure 14-16.

REAMERS

The *hand reamer* is a rotary cutting tool with teeth made on the outer surface of the body, Figure 14-17. It is used to enlarge previously drilled holes to accurate size and to make a hole smooth and straight. The parts of the hand reamer are shown in Figure 14-18. Hand reamers are identified by the square end of the shank. This shank is made to be driven with a suitable wrench using hand power. The best-style wrench to use is the T handle or tap wrench, Figure 14-19.

Machine power should never be used to drive a hand reamer. They are designed only to hand finish a hole to a high degree of accuracy and surface finish. Damage to the reamer will result because it is not made with enough clearances to be machine driven.

Also, a reamer should never be turned backwards at any time. This destroys the precision cutting edges. When a reamer is removed from a hole, it must be rotated in the same direction as it was cutting.

Hand reamers are made in many sizes and styles. Adjustable styles are made to slightly oversize a hole for more clearance to a mating part.

The *spiral-fluted reamer*, Figure 14-20, is a better design to eliminate chatter. The helical flutes bridge the gap and prevent binding when reaming holes that have keyways or grooves.

Taper pin reamers are made to ream fit an accurate hole to seat a taper pin, Figure

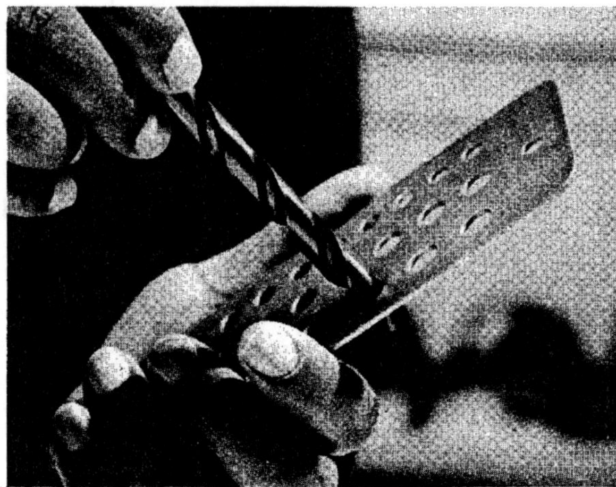

**Fig. 14-16 Using drill-size gage
to check twist drill size**

Fig. 14-17 Straight-fluted hand reamer

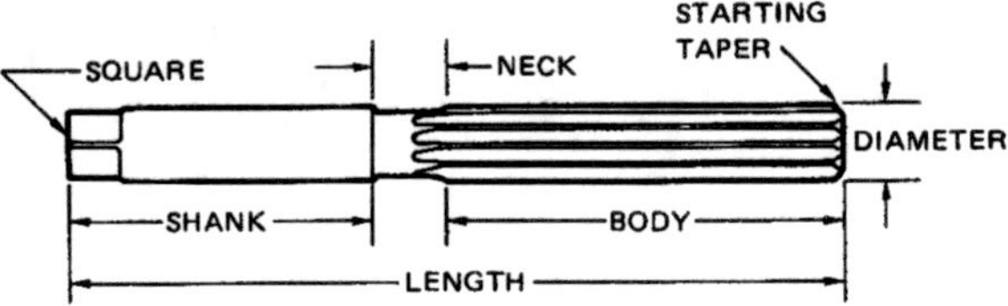

Fig. 14-18 Parts of the hand reamer

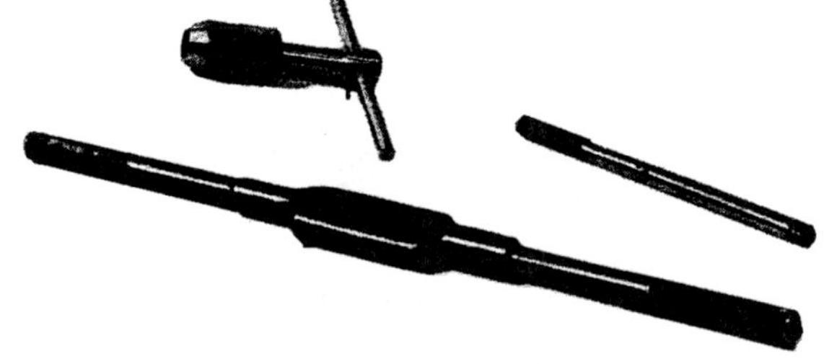

**Fig. 14-19 T handle (1) and tap wrench (2) are used
to drive the reamer (3)**

Fig. 14-20 Sprial-fluted reamer

Fig. 14-21A Straight-flute, taper pin reamer

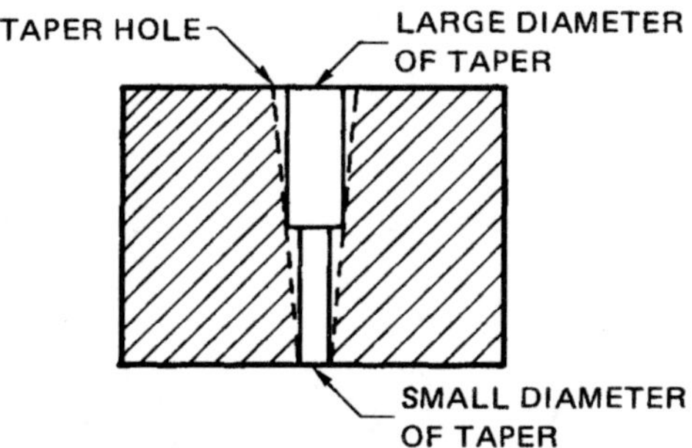

Fig. 14-21B The hole for taper reaming should be step drilled as shown

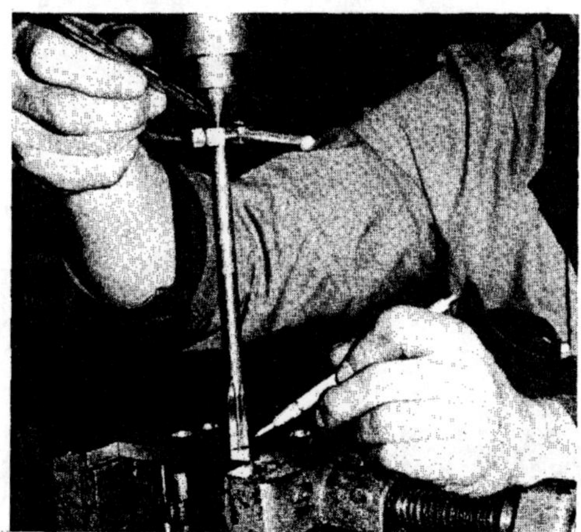

Fig. 14-22 Lining up the reamer to hand ream a hole accurately

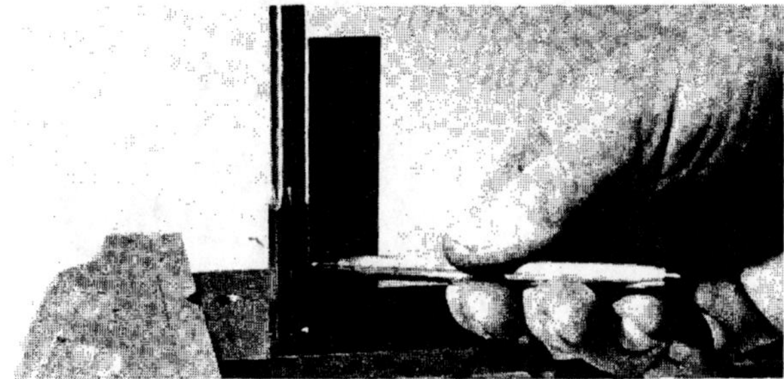

Fig. 14-23 Using the precision square to check the true alignment of the reamer

14-21A. Two holes slightly smaller than the reamer are drilled in the part before reaming. This step drilled hole makes the taper reaming easier, Figure 14-21B. Make sure enough material is left to produce the required taper reamed hole.

Reaming Procedure

A very small amount of material (0.002 inch to 0.005 inch) should be left to hand ream after the drilling operation. The hand reamer may be lined up to hand ream a hole accurately as shown in Figure 14-22. The hand reamer is made with a long, starting taper to provide alignment and straightness during the reaming process. Hand reamers do their cutting on this tapered section.

1. Check the drilled hole for correct size and surface finish so that it may be accurately hand reamed. A test hole may be drilled first and measured for correct size.
2. Lightly chamfer the hole with a countersink to remove burrs and provide a better alignment start for the reamer.
3. Select the correct-style and size reamer and check it for sharpness and accuracy.
4. Line up the reamer square to the hole to be reamed. Use the setup shown in Figure 14-22 or use a square at two 90-degree positions, Figure 14-23.
5. Use a T handle or tap wrench to turn the square end of the reamer.
6. Rotate the reamer slowly and evenly, keeping it straight. Allow the reamer to align itself with the hole to be reamed.
7. Use a steady, heavy downfeed — 1/4-inch feed per revolution for a 1/2-inch reamer and 1/2-inch feed per revolution for a 1-inch reamer.
8. Pass the reamer through the hole and remove it from the other side of the hole, if possible. If this cannot be done because of the part design, remove the reamer from the hole while continuing the forward rotation.
9. Use of a suitable cutting fluid improves the cutting action and surface finish when

reaming most metals. Cast iron and brass should be reamed dry.

10. Wipe the reamer clean and store it properly.

Reamer Care

- Reamers must be stored separately to protect their hard, sharp cutting edges.

- They should be kept in their original shipping tubes or placed separately in a tool stand.

- Never place reamers on materials which can dull or damage their accuracy.

BROACHES

The *broach* is a hardened and ground cutting tool, Figure 14-24. Metric broaches are also made. Hand broaches are used at the bench to produce standard internal keyways in hubs of gears, sprockets, and pulleys.

Bushings for various hole sizes are made in convenient sets, Figure 14-25. Precision shims are provided in the set to be placed behind the broach in the bushing slot to obtain the exact depth of a keyway.

Cutting the Keyway or Slot

1. Place the part to be broached on the table of the arbor press.

2. Make sure the hole is positioned over an open slot in the table plate to allow the broach tool to pass through the hole, Figure 14-26.

3. Place the correct size bushing in the finished bored hole of the part. Make sure a good grade of cutting oil is used while broaching. Cast iron is cut dry.

4. Select the correct-size broach from the set and insert it into the slot, in the bushing. Apply hand pressure to the arbor press handle, Figure 14-27. This causes the arbor ram to push the cutter broach into and through the part making a keyway or slot, Figure 14-27A.

5. After each pull on the arbor press handle, release the downward pressure on the

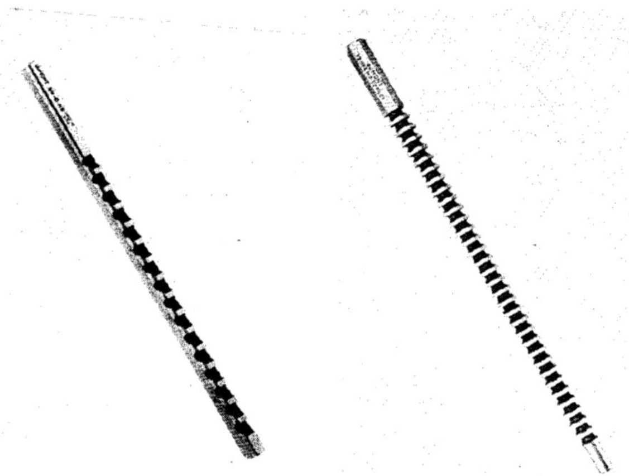

Fig. 14-24 Broaches (The Du Mont Corporation)

Fig. 14-25 Many bushings are available to fit different size holes that require broaching. (The Du Mont Corporation)

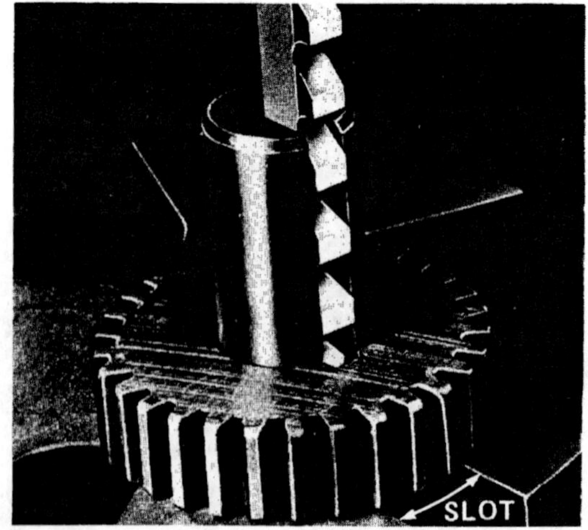

Fig. 14-26 Cutting broach is aligned to pass through slot in the table plate (The Du Mont Corporation)

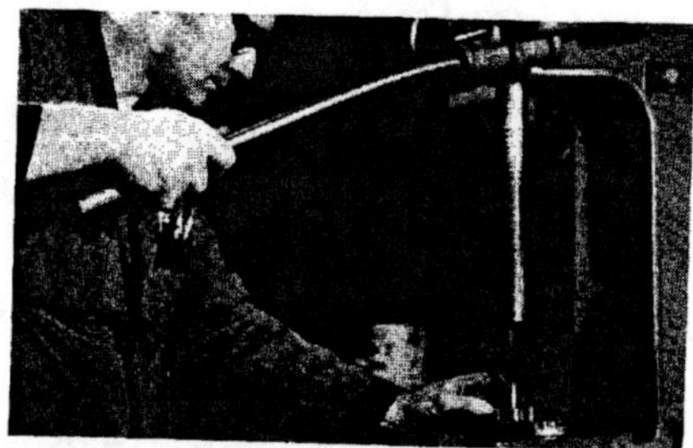

Fig. 14-27 Hand pressure is applied through the arbor press to force the cutting broach through the part.

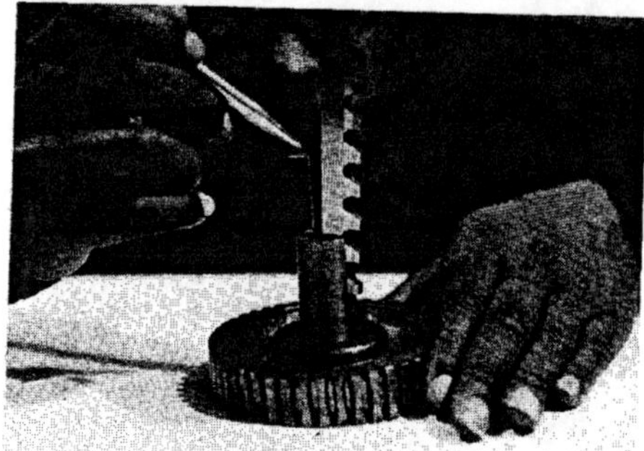

Fig. 14-27A Correct setup of gear, bushing, shim, and cutting broach. Note the placement of the shim to complete the keyway depth.

**Fig. 14-28 Finished keyway
(The Du Mont Corporation)**

broach. This allows the broach to reset itself square to the workpiece and not be in a bind. Binding of the hardened broach can cause breakage.

6. Place the required shim in the bushing slot behind the broach and make another cutting pass through the part, Figure 14-27. Additional cutting passes through the part with the use of shims of the correct thickness finishes the keyway to standard depth, Figure 14-28. Smaller broaches do not require shims. They are broached to depth in one cutting pass.

7. Remove the part from the arbor press. Clean and replace the bushing, shims, and cutter to their proper storage place.

8. With a smooth file, remove the cutting burrs from the edges of the keyway.

> **Caution:** Wear eye protection while broaching.

HAND SAWS

Hand Hacksaw

The *hand hacksaw* consists of a frame, handle, and cutting blade, Figure 14-29. The hacksaw is used to cut a wide variety of materials at the bench. These materials include metal sheets, metal bar, pipe, tubing, screws, bolts, brass, bronze, copper, nylon, and plastics.

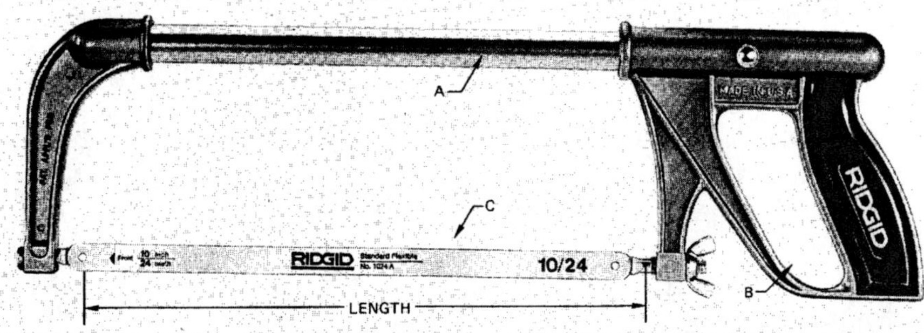

**Fig. 14-29 Parts of the hand hacksaw: (A) Frame, (B) handle, and (C) cutting blade
(The Ridge Tool Company)**

The hacksaw frame provides tension or tightness to the blade. The frame is made adjustable to use different blade lengths. A wing nut or rotating threaded handle adjusts the tightness or tension on the blade.

Be sure the blade is mounted with the teeth forward to cut on the forward stroke. Some blades are marked with arrows showing the correct mounting direction. Other blades have very fine starting teeth on the end of the blade away from the handle.

Hacksaw Blades

The hacksaw blade length is measured by the distance between the mounting holes in the blade, Figure 14-29. Hacksaw blades are made with flexible backs and hard teeth. Some are made with alloys and others with tungsten-carbide particles to cut glass and other hard materials.

The blades are made with different numbers of cutting teeth per inch, or pitch, for various sawing operations.

- 14 teeth – Used for most metals with thickness of 1 inch (25.4 mm) or more.

- 18 teeth – General-purpose blade for cutting steel bars, brass, and copper with 1/4-inch (6.3 mm) to 1-inch (25.4 mm) thickness.

- 24 teeth – For cutting pipe and tubing with 1/16-inch (1.5 mm) to 1/4-inch (6.3 mm) wall thickness.

- 32 teeth – For cutting thin metal shapes and tubing with a wall thickness up to 1/8 inch (3.1 mm) thick.

The rule for correct blade selection is to have at least 2 to 3 teeth resting or in contact with the work material at all times. This rule prevents the stripping of the cutting teeth from the blade. Fine tooth blades are used to cut thin materials, while coarse tooth blades are used to cut thicker stock.

Hand hacksaw blades are made in three styles of set as shown in Figure 14-30.

How to Use the Hacksaw

1. When using a hacksaw, always be sure the work is held rigidly in a vise on the bench.
2. The correct work height for easier sawing is elbow height. This allows the arms to move the saw back and forth without rocking. Move the body back and forth with each stroke. Have the feet apart for better control of body weight, Figure 14-31.

RAKER
Used with coarse-tooth blades to cut
solid sections in general-purpose sawing

WAVY
Used with finer tooth blades to cut thin
sections of tube and pipe and structural steel shapes

ALTERNATE
Used to cut thin, nonferrous materials,
thin sheet metal, tubing, and plastics

**Fig. 14-30 Standard styles
of set for hacksaw blade teeth**

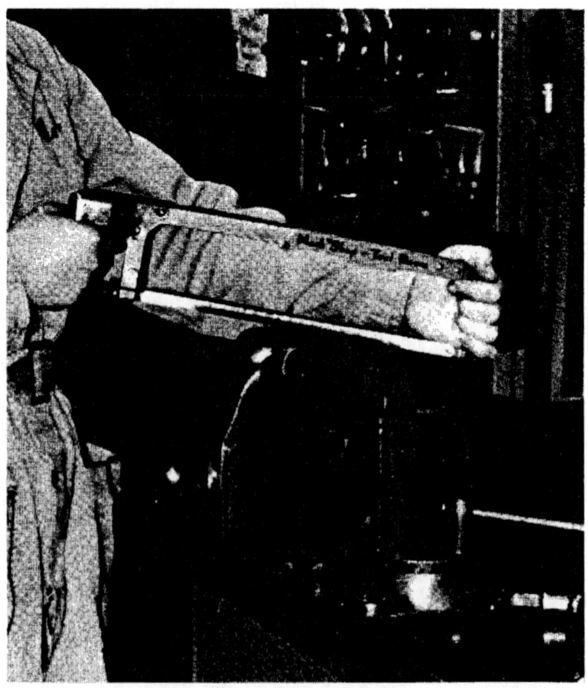

Fig. 14-31 Correct position to use hand hacksaw

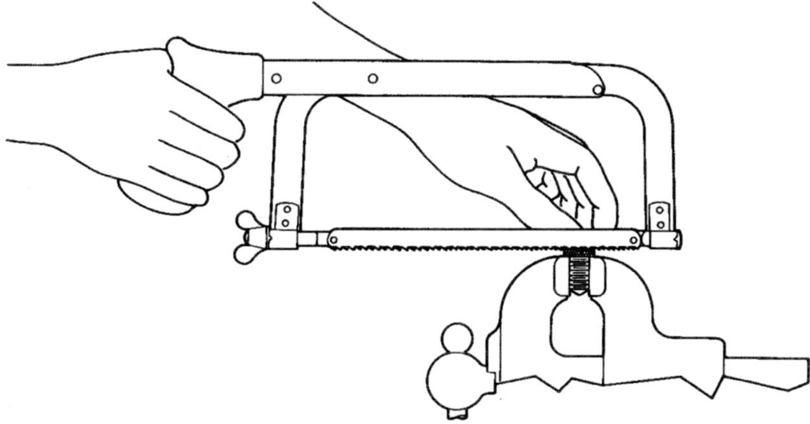

Fig. 14-32 Correct way to start the blade in work

3. Check the blade for the correct number of teeth, sharpness, and tightness.

4. A good cutting average for general-purpose sawing is 40 to 50 cutting strokes per minute.

5. To start cutting, use short strokes of from 1 inch (25.4 mm) to 2 inches (50.8 mm) using the thumb to guide the blade with a light, downward pressure on the forward stroke, Figure 14-32. No downward pressure should be applied on the return or back stroke. Avoid starting the cut on a corner or edge of a workpiece. Position the saw blade at a low angle to the work.

6. After the cut has been started, increase the length of the stroke from 6 inches (152.4 mm) to 7 inches (177.8 mm) to the full length of the blade.

7. Keep the saw cutting in a horizontal plane with strong, steady strokes directed away from yourself.

8. A workpiece can be nicked with the blade with a light, backward stroke. Harder metal can be nicked with the edge of a file to provide a good start for the saw.

9. Avoid twisting the blade which may cause it to break.

10. Use light cutting pressures as the saw nears the finish of the cut.

11. To saw thin wall tubing or pipe, it must be braced on the inside and held securely in the vise, Figure 14-33.

12. Hand hacksaw blades are usually used without the use of cutting fluids. Keep the blade from becoming excessively hot during use.

Hints for Sawing

1. Avoid starting a new blade in an existing saw cut. The existing cut has a narrower kerf than the width of the set of the new blade. The new teeth can easily be stripped off and dulled. It is best to rotate the work and begin a new cut.

2. Use less cutting strokes per minute with increased pressure when cutting hard, tough materials.

3. Thin materials may be sawed by clamping between two pieces of thicker material in the vise.

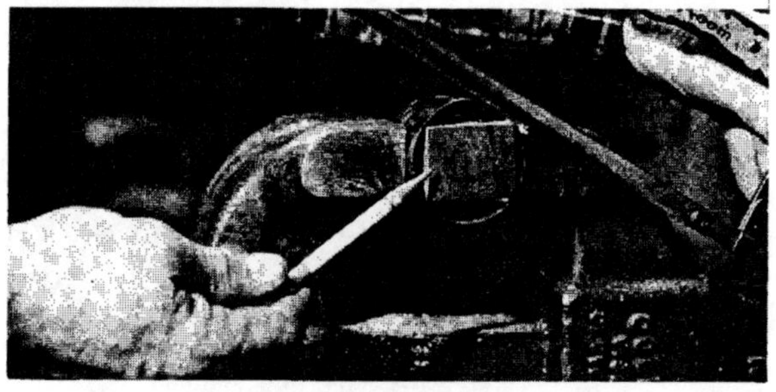

Fig. 14-33 To saw thin wall tubing or pipe, it must be braced on the inside and held securely in the vise.

4. Always use the coarsest tooth blade possible, having at least 2 to 3 teeth in contact with the work. The coarser pitch blade allows for faster cutting and better chip clearance.

Hole Saw

The *hole saw* is a circular, hardened, high-speed steel cutting edge welded to a tough alloy-steel back to cut round holes in materials. Assembled hole saws are shown in Figure 14-34. Hole saws are powered by portable, electric hand-drill motors or can be used in lathes or drill presses.

Figure 14-35 shows the threaded hole saw. The drill bit in the center cuts a hole in the material. The drill bit in the hole then serves as a pilot support for the saw to cut out the hole.

Hole sawing is an operation in which a core of material remains in the saw after the cut. Other cutting methods produce large amounts of chips. With hole sawing, the center part of the material may be saved and used for other parts.

Care of Hole Saws

- Hole saws should not be run too fast. Experience and feel aids the worker in selection of the correct speeds. Refer to the Appendix for a more comprehensive chart of *revolutions per minute* (rpm) for hole saws.

- Feeding pressure should be steady and at right angles to the work, Figure 14-36. Tilting or cutting at an angle may cause the saw to bind. This produces a deformed hole or strips off the teeth from the cutter.

- Certain materials, such as stainless steel and high-nickel metals, tend to work harden in the cut. Increased feed pressure must be applied so the teeth penetrate and cut. However, overfeeding causes the teeth to be ripped out of the saw body.

- Cutting fluids are recommended for longer life and cleaner cutting action. Cast iron is cut dry.

- Chips need to be removed either with cutting fluid or compressed air blast. If chips are not removed from the kerf at regular times, the saw quickly becomes dull by the abrasive action of the small recut chips.

Fig. 14-34 Assembled hole saws (Milwaukee Electric Tool Corporation)

Fig. 14-35 Threaded hole saw (Sioux Tools, Inc.)

Fig. 14-36 Apply correct feed pressure so the teeth cut through the material.

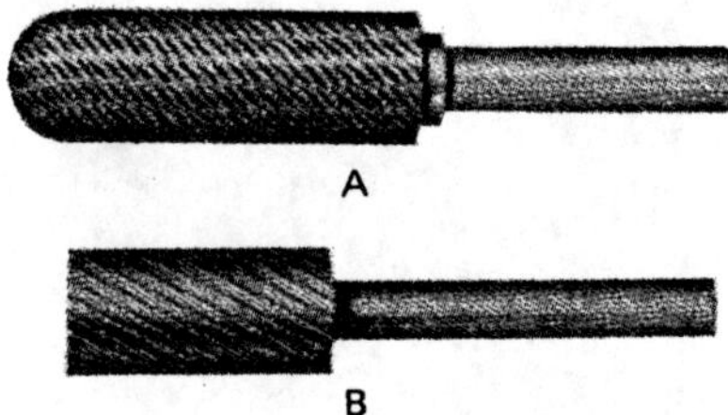

Fig. 14-37 Miscellaneous cutting tools: (A) Rotary power file, and (B) ground bur (Nicholson File Company)

Bur Diameter	Malleable Iron, Cast Iron, Tool Steels, Brass, Aluminum	Stainless Steels
1/16	50,000	75,000
3/32	40,000	60,000
1/8	35,000	53,000
3/16	25,000	38,000
1/4	22,000	33,000
5/16	20,000	30,000
3/8	18,000	27,000
7/16	17,000	26,000
1/2	16,000	24,000
5/8	15,000	23,000
3/4	14,000	21,000
7/8	13,000	20,000
1	12,000	18,000

Table 14-1 Speeds for carbide burs (Nicholson File Company)

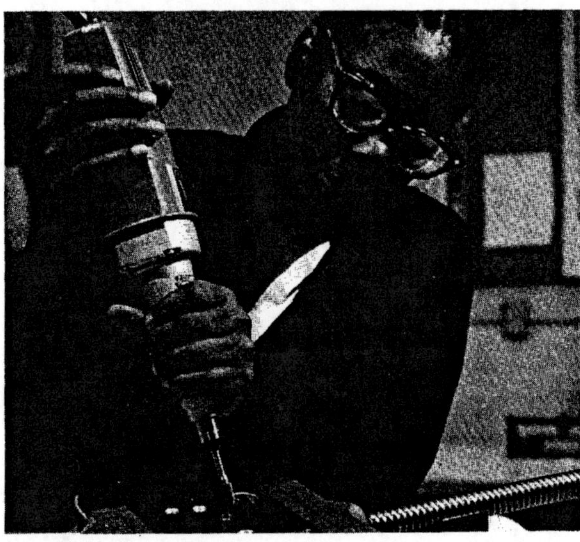

Fig. 14-38 Coarse-cut bur used for rapid stock removal

- Hole saw arbors are usually made with a 1/4-inch drill bit. Always use the arbor that matches the saw diameter to be used. For best driving grip, use the arbor with the largest size shank that fits the drill chuck being used.

- Use great care when starting a hole saw on an uneven surface as the teeth tend to grab and break.

OTHER CUTTING TOOLS

Rotary power files, or *ground burs*, are made in various styles for many purposes, Figure 14-37. The rotary file is widely used in die making, mold finishing, toolmaking, and finishing small and intricate parts. In such work, they are used for chamfering corners and removing burrs and sharp edges. Rotary power files are also used to make holes and slots longer in hard-to-reach places. This tool is a valuable time-saver at the bench.

How to Use the Rotary File

1. Mount the bur into a high-speed hand grinder.
2. Move the file or bur at an even rate and pressure to produce a good finish.
3. The speed of a bur depends on its size, Table 14-1.
4. Use only sharp files or burs.
5. When placing a bur in the grinder, use a short grip on the shank for accurate control. A longer grip on the shank may be used for hard-to-reach places using less cutting pressure.
6. Medium cut files or burs give sufficient stock removal and provide an acceptable finish. Special-cut files are made for soft materials. Coarse-cut tools should be used for heavier cutting and greater stock removal, Figure 14-38. Rotary power files and burs are also available in sets.

Caution: Always wear safety goggles and face protection to prevent face and eye injury from small, hot particles.

Metal cutting snips, Figure 14-39, are made to cut sheet metal, screening, wire, leather, gasket materials, rubber, plastics, and other commonly used shop materials. Other styles available besides straight snips are left-hand, right-hand, and heavy-duty shear snips.

Metal bolt cutters, Figure 14-40, are used to cut the ends or heads from bolts, pins, and rivets.

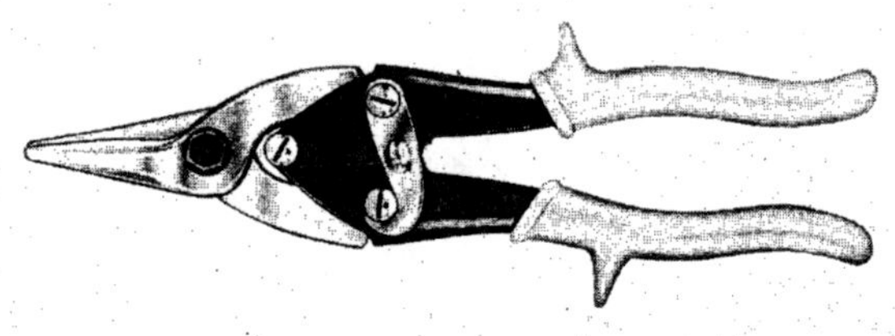

Fig. 14-39 Straight snips (Channellock, Inc.)

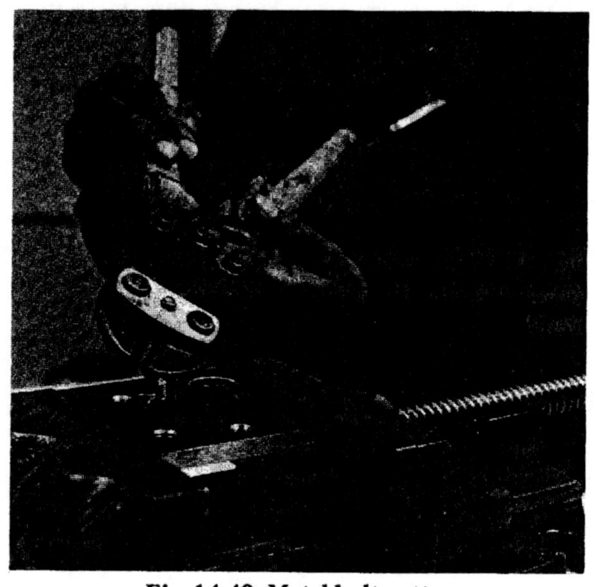

Fig. 14-40 Metal bolt cutters

REVIEW QUESTIONS

A.   Multiple Choice

1. Drilling is best described by:
   a. beveling an existing hole edge.
   b. producing a hole in material.
   c. enlarging the upper part of a hole.
   d. making an existing hole straight and true.

2. Reaming is best described by:
   a. beveling an existing hole edge.
   b. producing a hole in material.
   c. enlarging the upper part of a hole.
   d. making an existing hole straight and true.

3. ___________ removes the outer sharp edges of a hole and prepares the hole for tapping.
   a. Countersinking   c. Drilling
   b. Chamfering   d. Counterboring

4. Countersinks for tapered screw heads have an included angle of:
   a. 82 degrees.   c. 60 degrees.
   b. 70 degrees.   d. 120 degrees.

5. Twist drills for general-purpose drilling have an included angle of:
   a. 90 degrees.   c. 135 degrees.
   b. 118 degrees.   d. 187 degrees.

6. Which of the following amounts of material is correct to leave in a hole to be hand reamed?
   a. 0.001 inch to 0.005 inch   c. 0.006 inch to 0.018 inch
   b. 0.008 inch to 0.012 inch   d. 0.020 inch to 0.030 inch

7. Hand broaches are used to produce:
   a. tapered pin holes.
   b. a hole in a gear or sprocket.
   c. a keyway or slot in a hub.
   d. a keyseat in a shaft.

8. Which of the following tools is used to push broaches?
   a. Surface plate   c. Arbor press
   b. Bench vise   d. Bench plate

9. Broaches are fed to the correct depth with:
   a. rods.   c. shims.
   b. pins.   d. keys.

10. Which of the following hacksaw blade teeth per inch is used for general cutting?
    a. 14 teeth per inch   c. 24 teeth per inch
    b. 18 teeth per inch   d. 32 teeth per inch

11. Which of the following is the recommended cutting speed for hand hacksawing?
    a. 40 strokes to 50 strokes per minute
    b. 60 strokes to 70 strokes per minute
    c. 70 strokes to 80 strokes per minute
    d. 80 strokes to 90 strokes per minute

12. At least how many teeth of the saw blade should be in contact with the work at all times?
    a. 1 to 2 teeth   c. 3 to 4 teeth
    b. 2 to 3 teeth   d. 4 to 5 teeth

13. Which of the following tools cuts a hole in a part and retains the cut core?
    a. Rotary file   c. Broach
    b. Hole saw   d. Reamer

B.   Short Answer

14. List the two parts of a twist drill that must be equally sharpened to drill accurately.

15. How is the length of the hacksaw blade measured?

ACTIVITY

1. Demonstrate the correct use of the following hand tools:
   a. Reamer
   b. Hacksaw
   c. Keyway broach

2. Demonstrate the correct procedure to freehand sharpen a twist drill.

UNIT 15 TAPS AND DIES

OBJECTIVES

After completing this unit, the student will be able to

- identify the common taps, dies, and drive wrenches used in benchwork.

- describe and demonstrate the correct procedure to cut an inside and outside thread.

- demonstrate the proper use of formulas and tables to select the correct tap drill size for American National Standard screw threads.

INTRODUCTION

Threads are used in many mechanical assemblies to hold parts together. Threads offer great flexibility, ease, and speed in the assembly or disassembly of parts. Threads are made with precision cutting tools called *taps* and *dies*.

Taps and dies are designed to cut material away to produce a thread on the workpiece. As the tool turns, it cuts into the material. The material that is removed flows into the flutes of the tap or die. This leaves an evenly spaced helical groove on the part called a *thread*. The point of the tap and one end of the die are tapered to cause the cutting action. Also, the cutting edges are relieved to provide cutting clearance.

Hand taps and dies are made for standard thread forms in all standard sizes and pitches including metrics. *Pitch* is the distance from a point on one thread to a corresponding

point on the next thread. The pitch, or threads per inch, is easily measured with a thread pitch gage, Figure 15-1. Taps and dies are made of high-quality tool steel, hardened, and ground to size. The cutting edges are sharp and must be protected from damage by correct use and proper storage.

TAPS

The *hand tap*, Figure 15-2, is a small tool used for cutting threads on the inside of a hole, such as in a nut. One end of the tap is made square so that a tap wrench may be used to turn the tap into the hole. The squared end of the tap usually has a center hole. This can be used to guide the tap when started in a drill press or tapping fixture, Figure 15-3.

Tapping is one of the most difficult shop operations because of chip removal and lubrication to the cutting edges. Taps are easily broken because of their shape, hardness, and

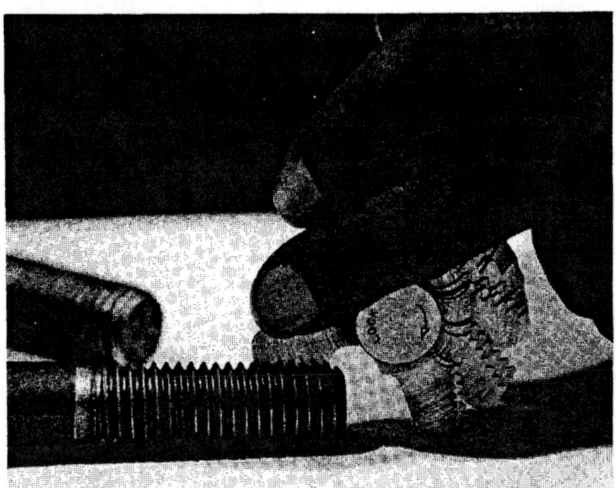

Fig. 15-1 Measuring pitch with a thread pitch gage

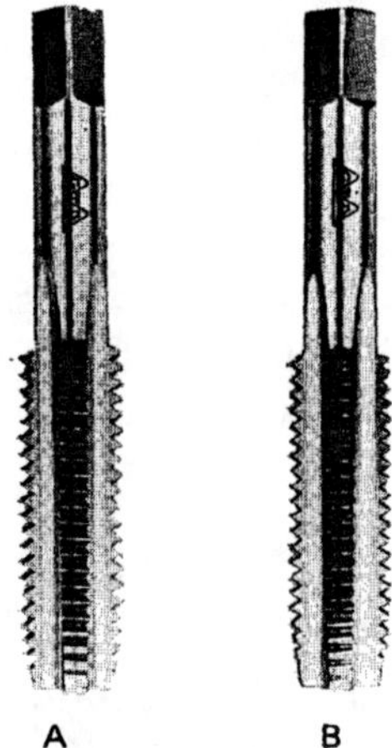

A   B

Fig. 15-2 (A) Left-hand tap, and (B) right-hand tap (Greenfield Tap & Die, Division of TRW, Inc.)

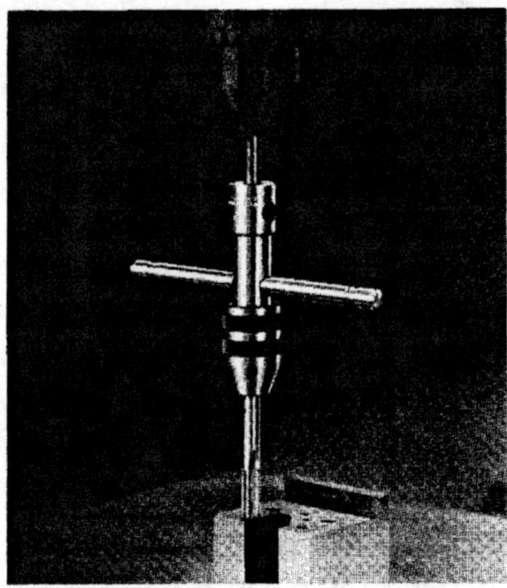

Fig. 15-3 Special tap wrench used to line up the tap square with the drilled hole in a drill press (Walton Company)

brittleness. Tapping holes accurately depends on the operator's skill, selecting the proper tap, drilling the correct hole size, and using the proper cutting fluid.

The cutting edges of the tap are formed by *flutes* (machined grooves) cut lengthwise through the threads. These flutes provide chip clearance and allow coolant to reach the cutting edges. The first few threads do all the actual cutting. The remaining threads guide the tap during the tapping operation.

Identification markings on the tap are explained below.

Example:

5/8″ –   11 UNC - GH3-HS

5/8″ =   Major diameter in inches

11   =   Number of threads per inch

UNC =   Unified National Coarse

GH3 =   Symbols used when tapping a specific size thread
G = Ground thread
H = Above basic size (an L would indicate below basic size)
3 = Pitch diameter limits

HS   =   High-speed steel (kind of material the tap is made from — usually this is HS or carbon steel.)

Machine screw taps are identified by similar markings.

Example:

12-28 UNF

12   = Gage size of the body of the screw

28   = Number of threads per inch

UNF = Unified National Fine

No designation is made for right-hand threads. The term LH is stamped if the thread is left hand.

Machine screw diameters under 1/4 inch are designated by gage numbers, such as #0, #4, #8, #10, and #12. Their actual major diameters are listed in the tables of tap drill sizes. The diameters of machine screws can be calculated easily by remembering that size #0 has a diameter of 0.060 inch. Each number size screw is then larger in diameter by 0.013 inch.

Types of Taps

Two types of taps commonly used in hand tapping are the standard tap and the spiral-pointed gun tap. The chips from the standard tap come up the flute space. The gun tap, Figure 15-4, is ground to shoot the chip out ahead of the tap, making it good for through hole tapping.

Hand taps are made in sets of three: taper, plug, and bottoming, Figure 15-5. The types of holes commonly tapped are open or through, blind but not bottoming, and blind bottoming, Figure 15-6.

The *taper*, or *starter tap* as it is often called, is tapered approximately 8 to 10 threads from the end making it easy starting. It can be used for starting threads in a blind hole or tapping holes which go through a workpiece. This style tap, with its long taper, also helps to center the tap in the hole.

The *plug tap* is tapered approximately 3 to 5 threads. This tap is the best first choice because it starts the thread, taps a blind hole deep enough, and taps through holes.

The *bottoming tap* is only beveled at the end for 1 1/2 to 2 threads. This tap produces threads to the bottom of a blind hole. Be careful to avoid breakage as the tap reaches the bottom of the hole. When tapping a blind hole, the sequence is as follows:

- Start with the taper tap.
- Use the plug tap next.
- Finish the thread to the bottom of the hole with the bottoming tap.

Number-size taps do not usually come in sets of three, as the finer pitches do not require a series of taps to thread a hole. Special kinds of taps are also made, such as left hand for cutting left-hand threads. Pipe taps are made with tapered threads to insure tight fits on pipe connections.

Fig. 15-4 Gun tap (Cleveland Twist Drill, An Acme-Cleveland Company)

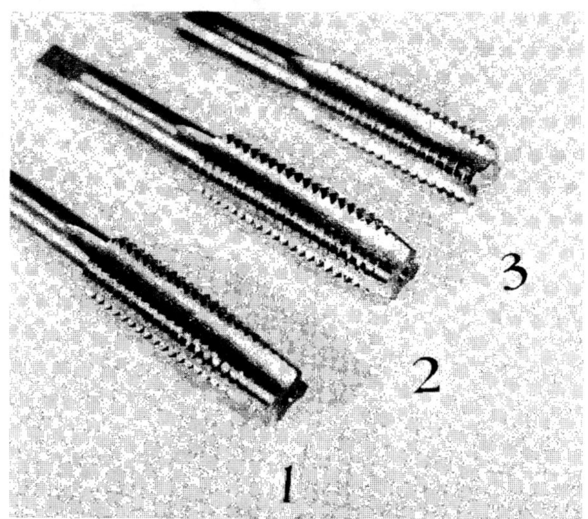

Fig. 15-5 Set of hand taps: (1) Taper, (2) plug, and (3) bottoming (Greenfield Tap & Die, Division of TRW, Inc.)

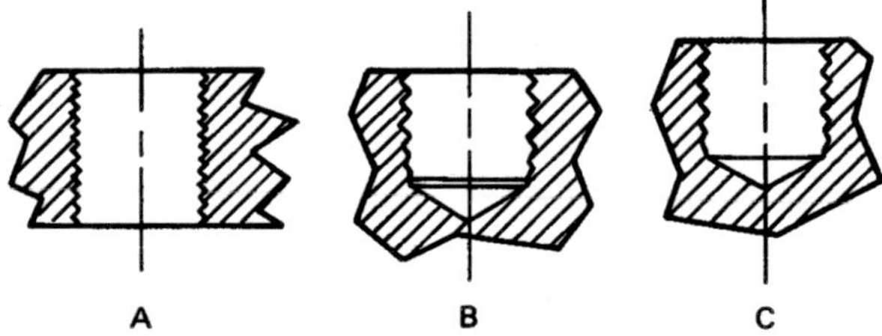

Fig. 15-6 Types of tapped holes: (A) Through hole, (B) blind hole, and (C) blind bottomed

Tap Sizes

Taps are available in all standard machine screw sizes from #0 to #80 to size 12, and from 1/4 inch up by 1/16 inch for standard screw sizes and pitches for the standard thread forms. Taps of special design and sizes for tapping special materials are available from manufacturers. Metric taps are also available and discussed in this unit.

How To Determine Correct Tap Drill Size

To produce an accurate thread in a workpiece, the correct size hole must first be drilled,

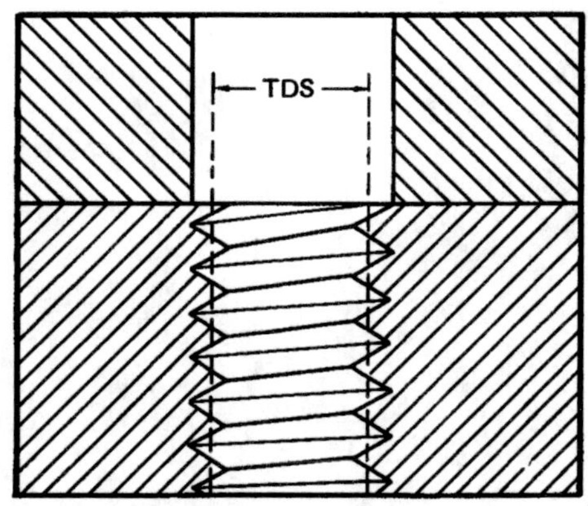

Fig. 15-7 The relation of tap drill size to a tapped hole

Figure 15-7. This correct drill size can be selected from available charts (see Appendix). Calculations may be made by use of the following simple formula to allow enough material to form about a 75 percent thread. The formula is:

$$TDS = D - \frac{1}{N}$$

TDS = Tap drill size

   D = Major diameter of tap

   N = Number of threads per inch

$\frac{1}{N}$ = Pitch of thread

The tap drill size is always smaller than the tap by the amount of the pitch $\frac{1}{N}$ subtracted from the major diameter (0).

Example: Find the tap drill size for a 3/4"-16 UNF tap.
        TDS = 3/4" - 1/16"
        TDS = 11/16"

Therefore, 11/16 inch is the correct tap drill size for a 3/4"-16 UNF tap. If the correct tap drill size is not available, always go to the nearest few thousandths larger standard drill size. Otherwise, the hole is drilled too small and tap breakage results. Use of slightly larger tap drills for the smaller sizes reduces tap breakage.

Thread depths over 75 percent give very little increased strength in most materials.

Common practice is to have a bolt engage a tapped hole by 1 to 1 1/2 times its diameter. The length of engagement must be enough to cause the bolt to break before the thread strips. Steel, cast iron, and brass must engage enough to withstand a shearing stress of 150,000 pounds per square inch (psi).

METRIC THREADS

Metric threads are produced in a workpiece in the same manner as American National Threads. These threads are cut with metric taps. In some cases, the tap holes are drilled with metric-size drill bits. This produces different percentage threads. Metric threads are available in coarse and fine series.

The International Standards Organization (ISO) Metric Thread Form is similar to the Unified Thread Form and is slowly being adapted by the United States. The System International (SI) Thread Form is similar to the American National Standard. Metric bolt sizes differ slightly from one European country to the next. Taps for the ISO metric threads are marked with the letter M and show the particular diameter size and the pitch.

Example: M6 × 1.0

      M = Metric
      6 = 6 mm diameter
      1 = 1.0 mm

To avoid any confusion, the metric sets of three tap types are defined as *taper tap, second tap,* and *bottoming tap.* In the United Kingdom, a bottoming tap is sometimes referred to as a *plug tap.* This is not the same as in the United States, where the plug tap corresponds with the British standard second tap.

Determining The Correct Tap Drill Size For Metric Taps

The correct tap drill size for metric taps may be obtained by use of the following formula. If available, first consult a table or chart (See Appendix). Using a chart gets the threading job done more quickly than by using a formula.

Hole size = outside diameter – (1.08254 × Pitch × Percent full thread)

Example: Find the correct tap drill size for the M6 × 1.0 tap for a 75 percent thread.

TDS = 6 mm – (1.08254 × 1.0 mm × 75%)
TDS = 5.0 mm

The tap drill size is again equal to the outside diameter minus the pitch. The correct pitch of the thread is measured with the metric screw pitch gage similar to the one in Figure 15-1 for American National Standard Threads. In the metric coarse series, the 12 common sizes range in diameter from 2 mm to 24 mm with a pitch range from 0.4 mm to 3.0 mm.

After the proper hole is drilled in the workpiece, it should be countersunk to remove the sharp hole edge left by the drill, Figure 15-8. This provides easier entry of the tap into the hole. Also, burrs thrown up on the hole edge by the tap above the level of the work surface are eliminated. These burrs are sharp and can cause injury to the worker. They can also cause misalignment in the assembly of parts. Burrs left on work are most certainly a sign of poor workmanship and should always be removed.

TAP WRENCHES

The adjustable, T-handle tap wrench and the double-ended adjustable tap wrench are used to hold the square end and drive the tap throughout the cutting operation. The *adjustable T-handle tap wrench* is used for confined areas, Figure 15-9A. The *double-ended adjustable tap wrench,* Figure 15-9B, is used when working on a surface which provides ample room. This tap wrench is used for larger taps. The longer handles provide additional torque (rotating power).

Great care must be taken to provide even torque to each handle to prevent tap breakage.

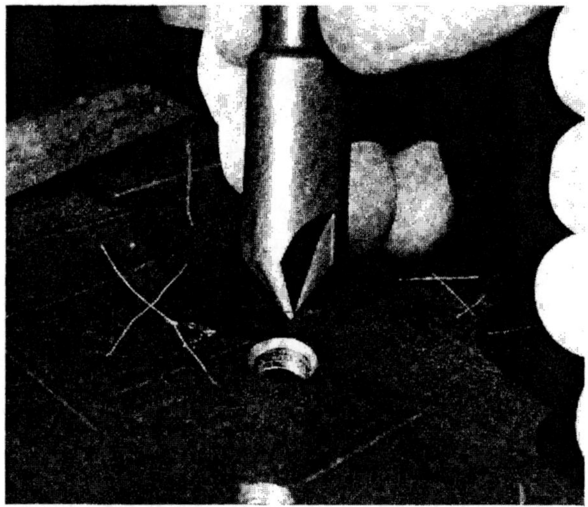

Fig. 15-8 Countersinking the hole before tapping produces a tapped hole free of edge burrs.

Fig. 15-9A Adjustable, T-handle tap wrench (Greenfield Tap & Die, Division of TRW, Inc)

Fig. 15-9B Double-ended adjustable tap wrench (Greenfield Tap & Die, Division of TRW, Inc.)

CUTTING FLUIDS

A proper cutting fluid must be applied during the cutting operation. Some materials, such as plastics, brass, and most cast iron, are cut dry. When required, cutting fluids provide the following:

- Smoother, easier cutting action
- Improved surface finishes
- Increased tap life
- Chip removal
- Decrease tap breakage
- Improved thread quality

Recommended cutting fluids are shown in Table 15-1.

MATERIAL	CUTTING FLUID
Aluminum	Kerosene & Lard Oil or Kerosene & Light Base Oil
Ampco Metal (Aluminum-Bronze) Grades 8-12; 16-18; 20-24	Soluble Oil
Bakelite	Dry
Brass	Soluble Oil or Light Base Oil
Cast Iron	Dry or Soluble Oil
Copper (99% Plus)	Mineral Oil with Lard or Light Base Oil
Fiber	Dry
Magnesium	Light Base Oil diluted with 50%-40% Kerosene
Malleable Iron	Soluble Oil or Sulphur Base Oil
Manganese Bronze	Mineral Oil with Lard or Light Base Oil
Monel Metal	Sulphur Base Oil
Naval Brass	Mineral Oil with Lard or Light Base Oil
Phosphor Bronze	Mineral Oil with Lard or Light Base Oil
Plastic Thermoplastic; Thermosetting, Rubber (Hard)	Dry or Air Jet
Steels: Free Machining AISI 1100 Series Low-Carbon (Up to 0.25%)	Soluble or Sulphur Base Oil
Medium Carbon Annealed (0.30%-60%)	Sulphur Base Oil
Heat Treated (0.30-60% Carbon) 225-283 Brinell Tool-High-Carbon & High-Speed Steel Stainless	Chlorinated Sulphur Base Oil
Titanium	
Tobin Bronze	Mineral Oil with Lard or Light Base Oil
Zinc Die Castings	Kerosene & Lard Oil

Table 15-1 Recommended cutting or tapping fluids

SURFACE COATINGS

Care must be taken to keep the tap cutting when tapping work-hardening materials such as stainless steel. For these operations, specially coated taps are used to increase tap life, productivity, and the quality of work. Listed below are a few of the commonly used coatings.

- Nitriding provides a very hard and wear-resistant surface to the tap.

- Flash chrome plating finishes are used to prevent chips from welding to the tap and to reduce friction between the tap and the work.

- Oxide treatment is a thin black coating used to reduce friction between the tap and the chips.

- Electrolized treatment produces a uniform layer of hard, dense, nonmagnetic alloy which gives great resistance to chip weld and reduces wear on cutting edges of the tap.

SPECIAL TAPS

Pulley taps are made with long shanks to reach through a pulley rim to tap threads in the hub of pulleys for oil cups and set screw holes. The diameter of the shank is the full diameter of the threads. *Extension taps* are used for reaching to deep areas to tap a hole. Deeper holes can also be tapped because of the smaller shank diameter.

Cold-forming taps have no cutting edges or conventional flutes. The threads on this tap form the threads in the hole by displacing the metal in an extrusion process. Threads produced with this tapping method are much stronger because metal is not cut away, but rather displaced. Work-hardened internal threads are cold formed in ductile metals such as copper, brass, lead, and leaded steels. No chips are produced to cause a chip problem. Up to 50 percent more torque is required than conventional tapping. Special charts available from tap suppliers, must be used to determine the correct tap drill size.

High shear taps are used for tapping aluminum. *Special spiral-fluted taps* are made with spiral flutes to draw the chips out of the hole or help the tap bridge a cavity, (i.e., keyway) along the path of the tap. This tap is very useful in tapping soft stringy materials, such as brass, aluminum, leaded steels, and copper.

Interrupted thread taps have every other tooth removed. These taps are recommended for use on soft and stringy materials. The removal of alternate teeth provides grooves that allow chips to escape. Thorough lubrication of the material ahead of the cutting teeth is another advantage. This tap also has advantages in tapping tough work-hardening metals because of the reduced tapping friction.

Serial taps are made in sets of three. Each tap is progressively larger in pitch diameter and is designed to remove part of the metal to produce the finished thread. Each tap is identified by having 1, 2, or 3 rings cut at the end of the shank near the square end. These taps are used to hand tap deep holes in tough metals. *Acme taps* are used to cut threads in parts for power transmission and strength applications. These threads are also easily tapped with acme roughing and finishing taps.

Pipe taps are made for all the standard pipe sizes, Figure 15-10. Recommended tap drill sizes are given for each thread size, Table 15-2. The proper depth to tap a standard pipe fitting is shown in Figure 15-11. Pipe threads taper 3/4-inch taper per foot. Accurate thread fits should have the hole taper reamed before tapping.

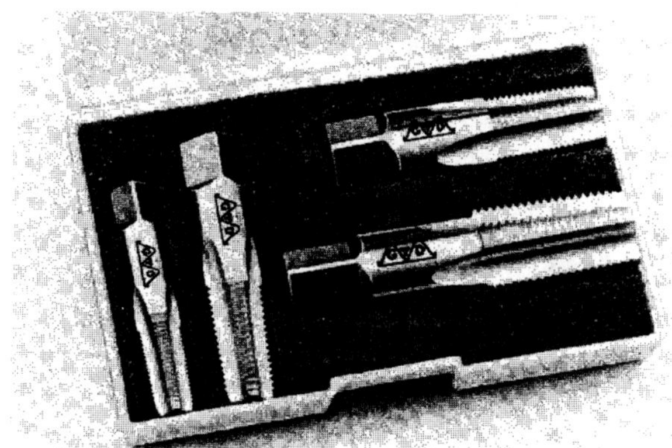

Fig. 15-10 Set of taper pipe taps (Greenfield Tap & Die, Division of TRW, Inc.)

PIPE THREAD SIZES			
Thread	**Drill**	**Thread**	**Drill**
1/8 - 27	R	1 1/2 - 11 1/2	1 47/64
1/4 - 18	7/16	2 - 11 1/2	2 7/32
3/8 - 18	37/64	2 1/2 - 8	2 5/8
1/2 - 14	23/32	3 - 8	3 1/4
3/4 - 14	59/64	3 1/2 - 8	3 3/4
1 - 11 1/2	1 5/32	4 - 8	4 1/4
1 1/4 - 11 1/2	1 1/2		

Table 15-2 Recommended tap drill sizes for each pipe thread size

Fig. 15-11 Part is tapped to the correct depth of the pipe tap

HOW TO HAND TAP THREADS IN A WORKPIECE

Tools and Equipment Required

- Tap or set of taps
- Tap wrench
- Tap drill
- Countersink
- Cutting fluid
- Bench vise with soft jaws to protect finished work
- Small steel square

Procedure

> **Caution:** Always wear safety goggles throughout the tapping operation.

1. Be sure the hole is drilled to the proper tap drill size, the correct depth and the ends are chamfered. When tap drilling holes larger than 1/2 inch (12.7 mm), first drill a hole 1/16 inch (1.5 mm) smaller than the finished tap drill size. (This hole aids the tap drill to produce the correct-size hole.)

2. Clamp the workpiece securely and squarely in a bench vise or on the bench. Place the hole in a vertical position, if possible.

3. Select a taper tap of the proper style and clamp the square end securely in a suitable tap wrench.

4. Apply proper cutting fluid to the tap, if required.

5. Grasp the tap wrench with the right hand cupped over the tap wrench and place the end of the tap in the hole in a vertical position, Figure 15-12.

6. Position the tap square with the top of the workpiece parallel with and at right angles to the hole.

7. Using the right hand with slight downward pressure, give the tap one full turn. Use a steady downward pressure to cause the tap to bite in and start the thread.

8. Steady the tap wrench with a light turning pressure of the left hand.

9. Check to make sure the tap has entered the hole and is cutting threads. Additional downward pressure may be required to cause the tap to start rather than tear or ream the top of the hole larger.

10. Check to make sure the tap is started squarely in the hole by sight. Make a final check for squareness by using a small square at two positions 90 degrees to each other, Figure 15-13.

Fig. 15-12 Proper method to start a tap cutting in the prepared hold

Fig. 15-13 Final check for squareness

11. A bench guide block, Figure 15-14, may be used to insure squareness of the tap with the hole.

12. Correct any misalignment of the tap by removing it from the hole and restarting it with pressure applied in the direction from which it leans. Be careful not to exert too much pressure at one time or the tap may break. Repeat this process and check until the tap is cutting squarely in the hole.

13. When the tap is correctly started, no further downward pressure is necessary as the tap draws or feeds itself into the hole. The tap only needs to be carefully turned with steady, even pressure to cut and produce the internal threads.

14. Reverse the tap about 1 turn after every 3 to 4 cutting turns, or when increased pressure is felt. This breaks the chip and provides tapping ease. This must be done carefully, with a steady motion, to clear the chips and prevent tap breakage.

15. Great care must be taken to apply the same turning pressure on both handles of the tap wrench and not to tip the wrench sideways. Also, never force a tap. It may become tightly wedged and may break off in the hole.

16. Continue the use of cutting fluid if required.

17. Proceed until the hole is finished tapped. Remove burrs and blow chips from the tapped hole.

> **Caution:** Always use eye protection when using compressed air to blow chips.

18. Check the tapped hole with a thread plug gage, Figure 15-15, for required accuracy.

19. Be careful when tapping into an existing round hole as in Figure 15-16. As the tap exits the hole into the side of a bored or drilled part, the tap is only cutting on two sides of the hole. This unbalance can cause tap breakage.

20. Save dull taps. They may be sharpened on a tool and cutter grinder by grinding

Fig. 15-14 Using a bench guide block to insure squareness of the tap with the hole

Fig. 15-15 Plug gage (Cleveland Twist Drill, an Acme-Cleveland Company)

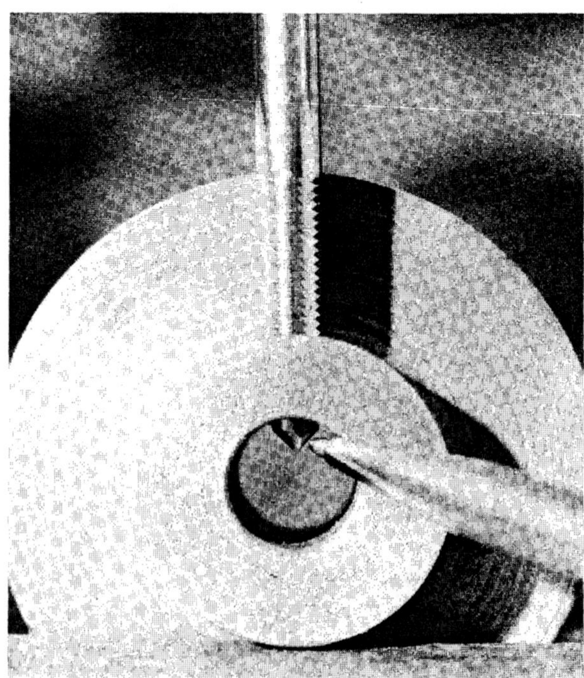

Fig. 15-16 Tapping into an existing round hole

through the flutes to sharpen the face of the cutting edges. A broken tap shank also makes an excellent center punch or lathe cutting tool when ground properly.

When tapping a blind hole, it is recommended that the beginner use all three taps of the set in proper sequence: taper, plug, and bottoming tap. Be sure to remove all chips from the hole with compressed air before using the bottoming tap.

As experience is gained, the plug tap can serve both as the starting and bottoming tap depending on the hole preparation. Great care must be taken in the tapping operation of any job. The tap is a brittle and fragile cutting tool and can easily be broken, especially in a blind hole. A broken tap can usually be removed, but sometimes it is a costly operation. If the broken tap cannot be removed, the workpiece must be scrapped.

If a tap is broken, consult your supervisor on the method of proper removal. Unit 19 discusses several methods of removing broken taps and studs.

THREADING DIES

Threading dies, Figure 15-17, are made the opposite of taps so they cut threads on the outside of bolts, rods, pipes, or other round cylindrical parts. The threads cut are called *external threads*. Dies are made in the same variety of materials, sizes, and thread forms as taps. The sizes of dies are designated by the same methods used for taps. Dies normally have four cutting edges and flutes. As with the tap, each flute provides an area for chip removal and lubrication during the threading operation.

The die has the threads tapered on one end to provide easier starting and squaring of the die to the workpiece. As the die is started on the workpiece, the taper or starting end of the die starts to cut a slight, not full, thread. Additional downward pressure and turning of the die on the workpiece causes the full thread to be cut. Once the die is started in a full cut thread, as with a tap, the die feeds itself along the workpiece as it is rotated.

TYPES OF THREADING DIES

There are five basic types of dies commonly used in shop benchwork.

The *round-split adjustable die* has a small screw used to adjust the opening of the die, Figure 15-17. With this adjustable feature, a thread can be cut with a roughing first cut and a finishing second cut. Another use of the adjustment is to thread a bolt to a desired fit into a tapped hole. *Split dies* differ in that there is no adjusting screw in the die. This die fits into a *stock*, or holder, and is adjusted by a tapered set screw in the die stock. This type of die must be reset each time it is changed in the holder.

Insert dies are a pair of matched dies used in an adjusting holding cap. The cap is assembled to a threaded collet head, locking it in position to a convenient guide section.

The *round, solid die, hexagonal, rethreading die*, and the *square nut die* are all used to refinish threads. They are not designed to cut a full thread. They should be used only to restore damaged, bruised, or rusty threads. The hexagon and square shapes make them convenient to use in close areas. (See Figure 19-2.)

DIE STOCK

The tool used for holding and turning the threading die and collet is called a *die*

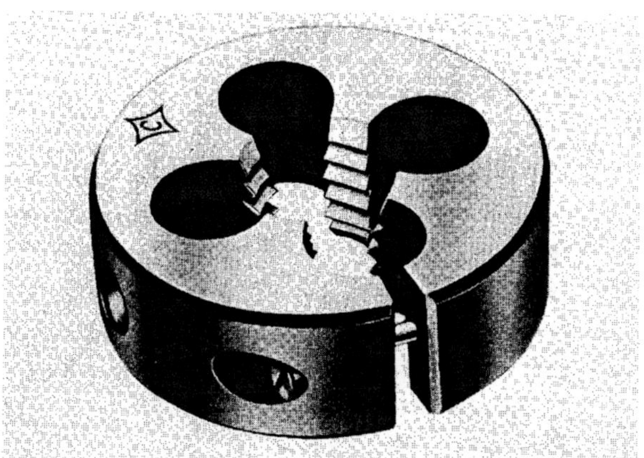

Fig. 15-17 Round-split adjustable die (Cleveland Twist Drill, and Acme-Cleveland Company)

Fig. 15-18 Die stock (Greenfield Tap & Die, Division of TRW, Inc.)

stock, Figure 15-18. It is usually just called a *stock*. Hand power is transferred through the two handles of the stock which are attached to a die holder. The single screw stock is designed to hold round dies. The set screw enters an indentation on the outside of the die. This screw holds the dies in the holder and provides drive to the die. Some die stocks have an adjustable guide built into the stock. This keeps the die in position with the work producing a straighter thread.

HOW TO CUT A THREAD WITH A DIE

Tools and Equipment Required

- Bench vise
- Small steel square
- File
- Threading die
- Die holder or stock
- Cutting fluid
- Screwdriver
- Nut or gage to test thread fit

Select stock the same diameter size as the desired threads. That is, 3/8″-16 UNC threads are cut on a 3/8-inch diameter shaft.

Procedure

1. Select the correct diameter material to be threaded. This should not be more than 0.005 inch (0.1270 mm) under nominal diameter and never oversize. Also, select the correct type of material called for on the print.
2. Bevel the end of the rod to be threaded by filing, grinding, or machine cutting to

Fig. 15-19 Bevel the end of the rod to be threaded

an included angle from 30 degrees to 60 degrees, Figure 15-19. This bevel allows the die to start easily and eliminates a burred first thread. This process is the same as chamfering a hole to be tapped.

3. Whenever possible, mount the work in a vertical position securely in a vise.
4. Select the correct die and mount it in the die stock. It is good practice to make a trial cut on a piece of scrap to be sure the setting is correct for the thread fit desired.
5. Apply the correct cutting fluid to the workpiece, if required.
6. Place the tapered or starting end of the die cutters against the beveled end of the rod to be threaded.
7. Hold the die squarely to the work.
8. Grasp the die stock with both hands near the die, Figure 15-20. Press down firmly while turning the die slowly until the first 2 or 3 threads are cut.
9. Once the die has started cutting, no more downward pressure is required. Rotating

Fig. 15-20 Proper way to hold die stock to start cutting

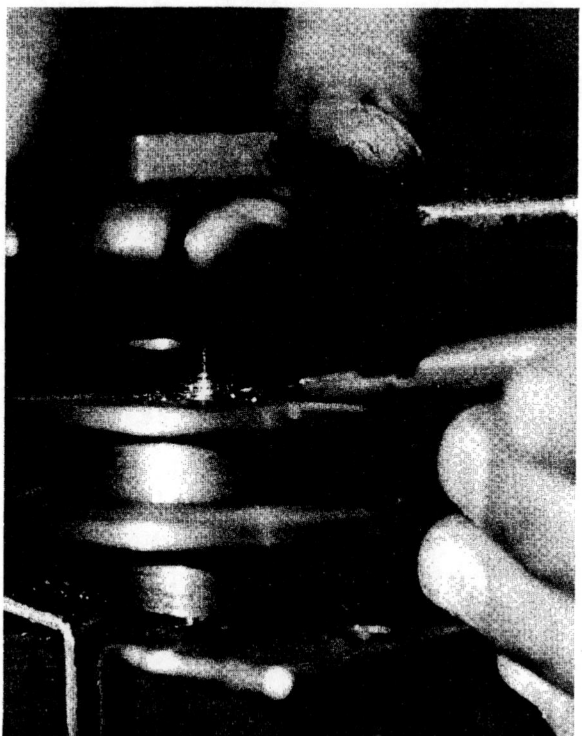

Fig. 15-21 Cutting a thread to a shoulder by turning the threading die over

the die stock causes the die to feed itself onto the rod and cut the thread.

10. With the small square, check at two places on the rod 90 degrees apart to be sure the die is started squarely. Use the same squaring procedure as described in tapping.

11. Use plenty of cutting fluid and continue threading.

12. Back off the die and check the thread fit with a gage or test nut. (Blow chips away and clean the thread before testing.)

13. Make necessary adjustments. Continue threading. Reverse the rotation of the die occasionally as resistance is felt to break the chip. This cleans the die, makes the threading easier, and produces a smoother thread.

14. When cutting threads to a shoulder, run the die down as far as possible, remove it from the work, and then turn it over, Figure 15-21. In this reversed position with the taper up, the die is fed down the rod and the threads are cut to the shoulder.

15. Clean the threads and remove sharp edges with a file.

16. Check the accuracy of the thread with a thread plug gage or with a thread ring gage, Figure 15-15.

CUTTING PIPE THREADS

The pipe thread is cut in a similar manner as general threading. Pipe threads have a taper of 3/4 inch per foot.

The workpiece being threaded should never be allowed to extend past the die end by more than one thread. This may cause the sealing effect of the thread to be destroyed. When the die is run past the workpiece end, the small part of the die begins to cut a straight thread instead of the tapered thread. See Figure 15-22 for correct position of the die and end of the part.

Ratchet stock wrenches are usually used to drive pipe ratchet dies. The wrench is designed to fit a set of dies. The normal range for a set of pipe ratchet dies is 1/16, 1/8, 1/4, 3/8, 1/2, 3/4, and 1 inch.

Other special tools, such as hand tap guides, are made to aid the benchworker in production tapping and threading. A power threading machine and a power tapping machine should always be used if available. The advantages of using power tapping and threading are increased speed, greater thread accuracy, improved finishes, and fewer broken taps and dies. Some workpieces, however, can only be hand tapped because of their design.

Fig. 15-22 A correct pipe thread is made when the die is even with the end of the part

REVIEW QUESTIONS

A.   Multiple Choice

1. A spiral-pointed tap is best used for tapping:
    a. blind holes.   c. through holes.
    b. bottomed holes.   d. UNF threads.

2. What is the next operation to the workpiece after the correct tap drill size hole is drilled?
    a. Tap   c. Thread
    b. Chamfer   d. Ream

3. Accurate threads are produced only when the tap or die is started:
    a. at an angle to the work.   c. beveled with the work.
    b. square with the work.   d. countersunk with the work.

4. External threads are hand cut with the use of a:
    a. tap.   c. thread pitch gage.
    b. die.   d. countersink.

B.   Short answer

5. Briefly define *tap drill size*.

6. State the nearest tap drill size to use for 75 percent full thread.
    a. 1/2" – 13 UNC Tap   TDS = ____________

    b. 9/16" – 18 UNF Tap   TDS = ____________

    c. 3/4" – 10 UNC Tap   TDS = ____________

7. Briefly explain the procedure for correcting a tap which has not started squarely.

8. Briefly state how to break the chip while tapping a hole.

9. Name the two types of wrenches used to hold and drive taps.

10. List four common causes of tap breakage.

11. State what must be done to the end of a rod before threading.

12. State the name of the tool to hold and drive the threading die.

13. List three styles of threading dies.

ACTIVITIES

1. From the working drawing, select the correct material to be used. Obtain suitable material from your instructor or material area. Proceed to prepare your material using correct layout procedures and tools. Select the correct tap drill sizes, the correct tap, and correct lubricants. Proceed to drill, chamfer, and hand tap the holes following the step-by-step procedure described in the unit.

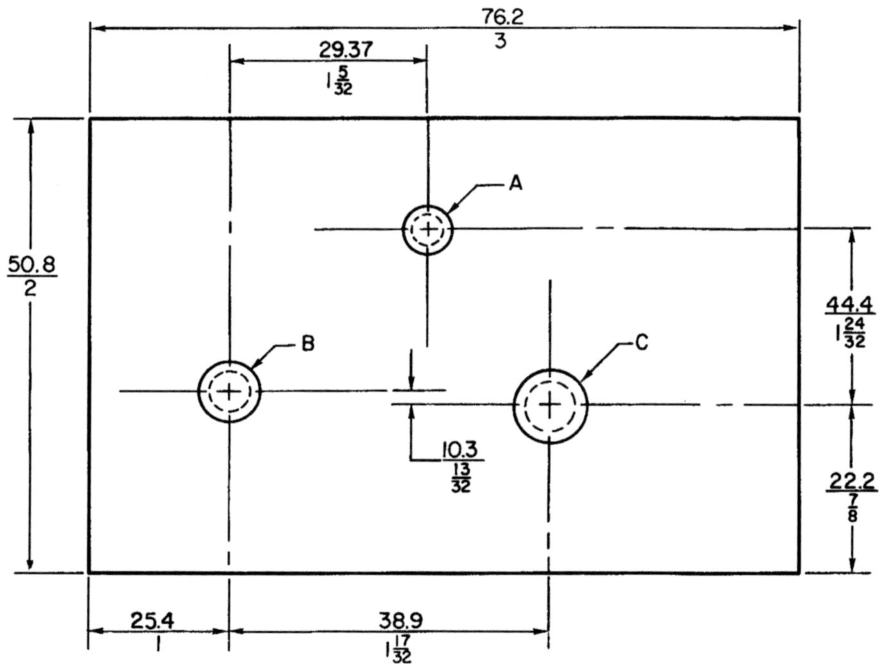

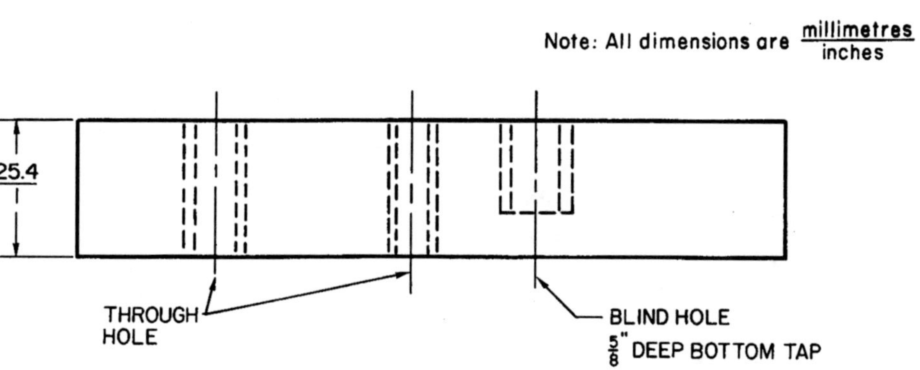

Material is mild steel.

a. Hole A = 1/4″ – 20 UNC – 2B   Tap Drill Size = __________

b. Hole B = 5/16″ – 18 UNC – 2B   Tap Drill Size = __________

c. Hole C = 3/8″ – 24 UNF – 2B   Tap Drill Size = __________

2. Hand thread the following:
   a. 1/4-inch diameter rod — 20 threads per inch
   b. 3/8-inch diameter rod — 16 threads per inch
   c. 1/2-inch diameter rod — 13 threads per inch

Use mild steel material provided by your instructor. Properly prepare, thread, and finish each rod.

SECTION V
POWER TOOLS

UNIT 16 POWER HAND DRILLS AND GRINDERS

OBJECTIVES

After completing this unit, the student will be able to

- identify the common hand-held bench power tools.

- define and demonstrate the safe and proper use of hand-held bench power tools.

INTRODUCTION

Many tools used in benchwork are powered beyond the hand power of the worker. Electrical and air-powered tools decrease the time and effort to complete a particular job. They also provide for greater accuracy in the finished workpiece. In today's high labor-cost industries, power tools are necessary to increase profits and productivity. Power tools are expensive. They are also capable of causing severe injury. Always use power tools wisely and safely to prevent injury and damage.

AIR-POWERED TOOLS

Air-powered tools are driven by an air motor built into the tool. The air is supplied to the turbine motor by an air compressor through a suitable control valve and hose. Air powered tools run cooler than electrical tools. They also are safer because there is no chance of electrical shock or sparks. Clean, dry filtered air must be used. Air tools must be properly lubricated and cleaned to prevent rust and corrosion to the internal parts. Air-powered tools can be stalled without damage to the tool.

ELECTRICAL-POWERED TOOLS

Electrical-powered tools require additional regular maintenance to the parts of the electrical motor. The motor bearings and gear section must be properly lubricated. The brushes and armature must also be kept in good order. The motor is air cooled and must be kept free of dust and dirt to prevent overheating. The electrical motor can burn out if stalled for an extended time. Use the manufacturer's recommended procedures for care and maintenance of all powered tools.

POWER HAND DRILLS

The *power hand drill* is probably the most often-used rotary power tool. The standard pistol-grip type is the most popular design because of the variety of areas in which it can be used with relative ease, Figure 16-1. Power hand drills are made in various sizes to drill holes from 1/16 inch (1.5 mm) to 1 inch (25.4 mm) in diameter. Special drills for smaller holes are available.

The sizes of portable, electric hand drills are classified by the maximum size, straight-shank drill it holds, such as 1/4, 3/8, and 1/2 inch.

Example: A 1/4-inch drill holds a drill bit up to and including 1/4 inch.

Light-duty work is normally done with the smaller drills having faster speeds. Larger, heavier drilling is done with the larger drills at slower speeds. The heavier drills are equipped with an additional rear and side handle to provide extra holding power and greater drilling pressure, Figure 16-2. Electromagnetic drill bases are provided for the heavier, larger drilling jobs, Figure 16-3. The base magnet attaches to the material to be drilled. The drill is fed by a hand lever.

Power hand drills are also made with 45-degree and 90-degree head angles to drill holes in close areas, Figure 16-4. The drill bit is held securely in a three-jaw device called a *drill chuck*. This drill chuck has a T handle wrench with a thumb pad called the chuck key. The chuck key inserts into the side of the drill chuck to tighten or loosen the drill bit in the hand drill, Figure 16-5.

> **Caution:** The chuck key should never be left in the drill chuck when turning on the drill. The key can be thrown out by the rotating chuck and cause severe personal injury.

> **Caution:** Be mindful of loose clothing, long, loose hair, fingers, and hands when power hand drilling. The drill bit is rotating and sharp and can cause personal injury.

Most drills today are made with a variable-speed control switch. This feature gives an infinite speed range for drilling various kinds of hard and soft materials.

Fig. 16-1 Pistol-grip hand drill (Skil Corporation)

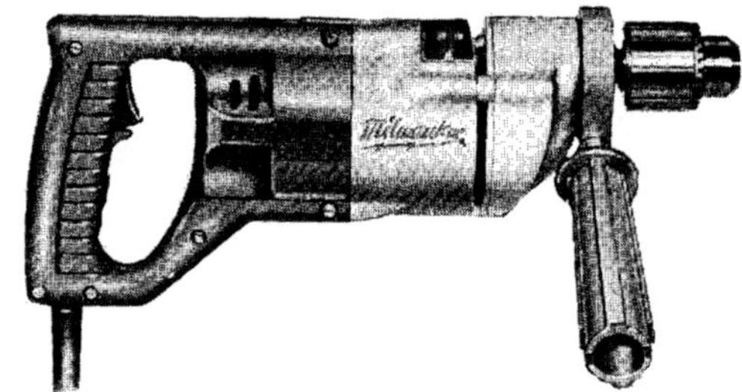

Fig. 16-2 Heavy, large drill equipped with a rear and side handle for extra holding power (Milwaukee Electric Tool Corporation)

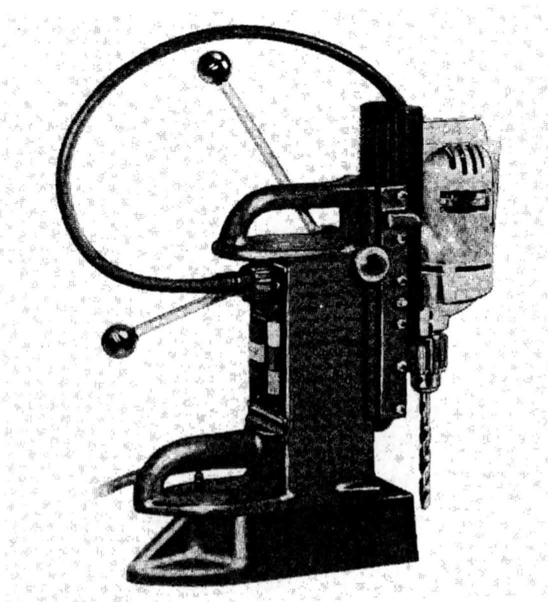

Fig. 16-3 Electromagnetic drill base (Milwaukee Electric Tool Corporation)

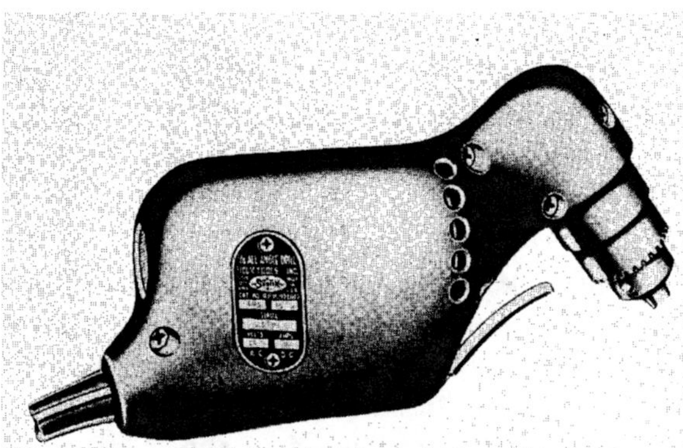

Fig. 16-4 Power hand drill with head angle (Sioux Tools, Inc.)

Fig. 16-5 Use of the chuck key to loosen or tighten the drill bit in the hand drill

Reversing power drills are available in all drill sizes. The reversing drill with driving bits can tighten or loosen threaded fasteners. Also, the drill can be reversed and the bit unscrewed if it becomes jammed in the workpiece. A lock button is provided so the trigger can be locked in an ON position. This frees both hands to apply additional drilling pressure. Squeezing and releasing the trigger releases the lock and shuts off the drill.

Power Hand-Drilling Procedure

1. Lay out the hole to be drilled.
2. Prick punch the intersecting centerlines and recheck the layout.
3. Center punch in the prick punch hole.
4. Center drill and pilot or predrill.
5. Select the correct drill bit size, making sure it is properly sharpened.
6. Place the drill bit in the correct-size power drill. Large-diameter bits used in smaller power drills can cause overheating and damage to an electric motor.
7. Drill the hole, starting the drill bit slowly in the center punch mark or predrilled hole.
8. Apply pushing pressure to the pistol grip of the power drill causing the drill bit to cut.
9. As soon as the full diameter of the drill bit is cutting, remove the drill bit from the hole and measure the diameter for accuracy.
10. Reduce the pushing pressure and the speed as the drill bit begins to exit the workpiece. This allows the drill bit to ease and cut through the material without becoming jammed.
11. Remove burrs from both ends of the hole with the countersink.
12. Harder metals are drilled best with increased pressure and slower speeds. To maintain this slow speed on power drills without variable-speed control, jog the trigger switch on and off. This allows the drilling speed to slow or drop off to the required slower speed.
13. Thin metal may be drilled by placing it between two thicker pieces.

Safety Precautions

- Prevent the power drill from twisting out of the hands. Always keep a good hold on the tool.
- Good footing and balance are also very important to prevent injuries.
- Power drills of 1/2 inch and larger should be held by two workers. The twisting torque is high in the larger power drills.
- Make sure the electrical cord is not damaged. Also, check the plug end to make sure the ground connection is in good condition.

Drill Speeds

Proper speed and feed of a drill is determined by the drill material, diameter of the drill, and the type of material to be drilled. Drilling speeds should be obtained from a chart or table (see Appendix). If these are not available, the following formula is used to determine the approximate drilling rpm.

$$\text{RPM} = \frac{4\,\text{CS}}{\text{DIA}}$$

RPM = Revolutions per minute of the drill bit

CS = Recommended cutting speed of the material to be drilled

DIA = Diameter of the drill bit

Note: CS is the distance a point on the drill circumference travels in surface feet per minute.

Example: Calculate the approximate rpm to drill a 1/2-inch (12.7 mm) hole in a piece of machine steel.

$$\text{RPM} = \frac{4\,\text{CS}}{\text{DIA}}$$

$$\text{RPM} = \frac{4 \times 100}{1/2}$$

$$\text{RPM} = 800$$

The recommended cutting speed (CS) for common materials is as follows:

Material	Cutting Speed
Steel Casting	40 feet per minute
Tool Steel	60 feet per minute
Cast Iron	80 feet per minute
Machine Steel	100 feet per minute
Brass and Aluminum	200 feet per minute

Note: These cutting speeds are recommended for high-speed steel cutting tools. For carbonsteel tools, use 1/2 the recommended cutting speed. These cutting speeds are to be used only as a starting point. Use enough drilling feed to keep the drill cutting clean. Too heavy a feed breaks the drill while too light a feed causes chatter and quickly dulls the cutting edges.

Jigsaw

The *jigsaw* is a hand-held, reciprocating saw-blade tool made similar in style to the power drill, Figure 16-6. Sets of assorted saw blades are available for cutting various materials. The jigsaw is used to power saw various materials. A drilled hole provides a convenient starting place for the jigsaw to saw square, circular, or irregular cutouts in materials. Thin metal may be sawed by clamping it between two pieces of plywood or other suitable material.

When using this power tool, follow the operating instructions provided by the manufacturer. Before changing blades, remove the electrical plug from the power outlet.

Hand Grinder

The *hand grinder* is another hand-powered tool that is often used at the bench. This tool grinds and sands flat surfaces and smooths welded areas. The wire brush attachment removes paint and rust. Other attachments include buffers and polishers.

Many wheel styles and types are made for various operations. Figure 16-7 shows the

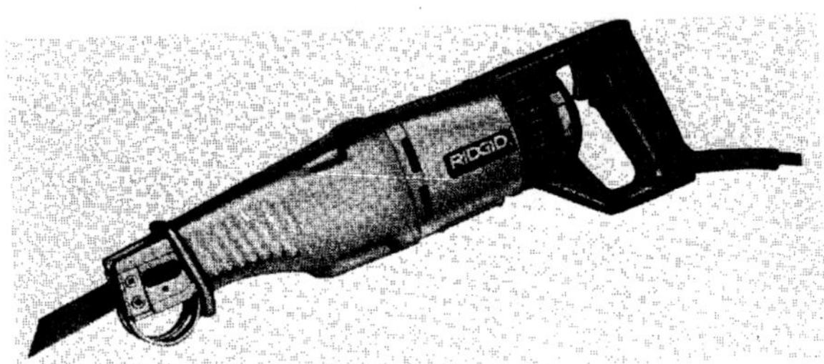

Fig. 16-6 Electric-powered, hand-held jigsaw (The Ridge Tool Company)

Fig. 16-7 Electric-powered, angle-head grinder (Skil Corporation)

electric-offset or angle-head grinder. Also available is an air-powered, angle-head grinder. The straight-style hand grinder is shown in Figure 16-8A. The portable, straight grinder is shown in Figure 16-8B. The nose of the hand grinder has a threaded spindle to which a sanding disc back plate can be screwed. Figure 16-9 shows the sanding disc installed on the sander.

A high-speed die grinder is used to grind hard-to-reach areas, especially in the making of dies, Figure 16-10. Thin, small abrasive wheels may be mounted to a special arbor on these tools to abrasive cut materials. Various-shaped cutouts in thin metal may also be done with abrasive cutting. Small, mounted grinding wheels are often used to grind various shapes in a workpiece. Air-powered die grinders are also available.

Hand grinder speeds range from 5,000 rpm to 10,000 rpm. The size of the hand grinder is determined by the diameter of the grinding wheel it is designed to hold. Thus, a 6-inch grinder uses a 6-inch wheel.

Grinding Hints and Safety Precautions

- When using the angle-head grinder, faster cutting is done with the edge of the wheel being pushed into the workpiece, Figure 16-11. The flat surface of the wheel is used for leveling and finishing operations.

- Protect the wheel from damage. A crack can cause the wheel to fly apart.

- Power wire brushing should be done with great care as the wire bristles tend to pull and move the workpiece. Always clamp the work securely. Always start

Fig. 16-8A Straight-style hand grinder (Milwaukee Electric Tool Corporation)

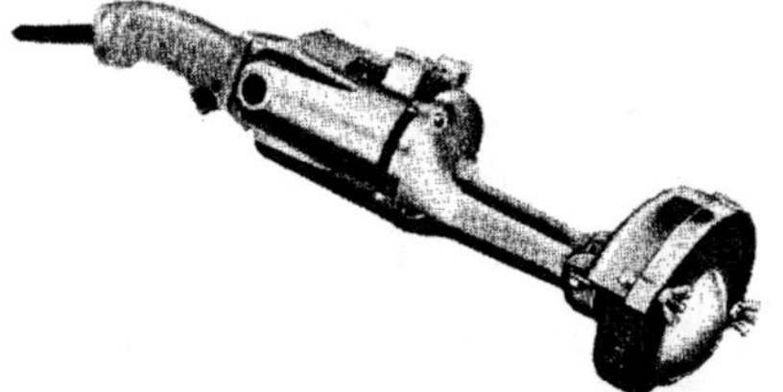

Fig. 16-8B Portable, straight grinder (Milwaukee Electric Tool Corporation)

Fig. 16-9 Sanding disc installed on the sander (Sioux Tools, Inc.)

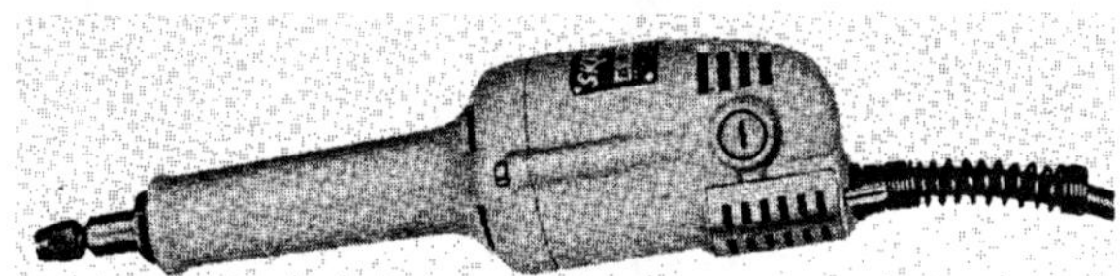

Fig. 16-10 Small, high-speed die grinder (Skil Corporation)

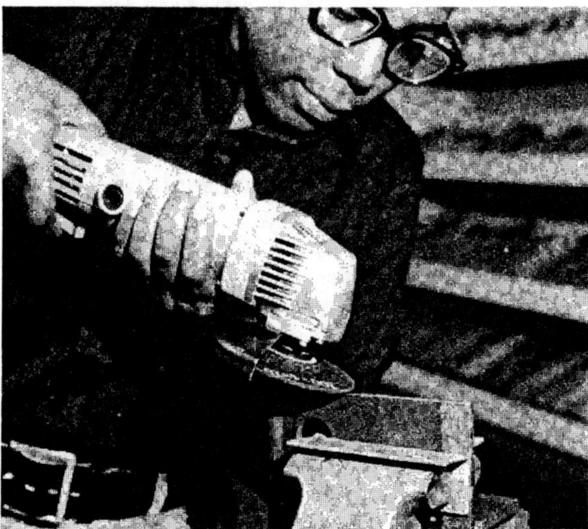

Fig. 16-11 Using the outside edge of the angle-head grinder wheel for faster material cutting

the grinder before contacting the work-piece.

- At times, the grinder may be clamped in a vise and the work held by hand to the grinder or wire brush. Extreme care must be taken so the workpiece is not pulled out of the hands.

- Care also must be taken not to get gloves or fingers into the rotating wheel.

- Wire bristles can loosen and fly out of the wheel. Eye and face protection must be worn when wire brushing or grinding.

- Hold the grinder with two hands; one on the head handle to provide pressure to the operation and the other on the trigger handle to guide and hold the tool.

- Always keep wheel guards in place.

- Always wear safety glasses or a face shield when grinding.

- An abrasive-cloth belt sander can also be used to remove burrs. With a belt sander, workpieces are finished quickly and easily, Figure 16-12.

Fig. 16-12 An abrasive belt sander is extremely useful as a finishing tool

REVIEW QUESTIONS

A. Multiple Choice

1. An electric-powered drill can easily be damaged by:
   a. drilling brass.
   b. drilling steel.
   c. using too small a drill bit.
   d. using too large a drill bit.

2. When drilling a hole, as the drill breaks through the material, which of the following should be done?
   a. Increase feed
   b. Decrease feed
   c. Tip the drill
   d. Increase the speed

3. The approximate cutting speed to drill brass using a high-speed, steel drill bit is:
   a. 200 sfpm.
   b. 80 sfpm.
   c. 40 sfpm.
   d. 600 sfpm.

4. A hand grinder is sized by the:
   a. speed of the grinder.
   b. diameter of the wheel it is designed to hold.
   c. length of the handle.
   d. horsepower of the motor.

B.   Short Answer

5. State three benefits of using powered hand tools.

6. Name the two sources of power used to drive hand tools.

7. Name the source of power of the tool that is not damaged when stalled.

8. State the best feed and speed combination to drill harder metals.

9. State the reason why it is important to hold on to a drill with both hands.

10. State two benefits of the reversing feature on a powered hand drill.

11. Using the rpm formula, $RPM = \dfrac{4\ CS}{DIA}$, calculate the approximate rpm to drill a 1/4-inch hole in machine steel. Show your work.

12. List five safety precautions when using powered hand tools.

ACTIVITY

1. Identify each of the following powered hand tools and safely demonstrate the correct use of each.
    a. Piston-grip drill
    b. Heavy-duty drill
    c. Angle-head grinder
    d. Angle-head sander
    e. Straight and portable grinders
    f. Jigsaw
    g. Die grinder

UNIT 17 POWER SAWS

OBJECTIVES

After completing this unit, the student will be able to

- identify the common metal-cutting band saws used in benchwork.
- demonstrate the correct safe use of the band saw.
- select and install the correct band saw blade for a job.

INTRODUCTION

There are several different styles of power saws available to cut off metal and other shop materials. The first operation to be done on a workpiece is to cut it to the required length. Band saw machines are fast because of the narrow cuts made.

POWER SAWS

The *reciprocating power hacksaw* uses a back-and-forth motion to cut material, Figure 17-1. This machine is a powered, heavy-duty, hand hacksaw held in a rigid frame. The saw blades are heavier than those used with the hand hacksaw. The same blade selection and use apply to the reciprocating power hacksaw as the hand hacksaw discussed in Unit 14. A disadvantage of this power saw is that it cuts only on the forward stroke with a wasted return stroke.

The reciprocating saws are rapidly being replaced by the metal-cutting band saw. The *metal-cutting band saw,* Figure 17-2, uses a continuous band blade with teeth on one edge. The endless band blade cuts at all times with no loss of time. The band saw has the advantage of cutting shapes rapidly, accurately, safely, and easily. It also saves material and is inexpensive to use. All these advantages result from the band machine removing material in sections rather than in wasted chips.

BAND SAW BLADES

Band saw blades are available in ready-cut and welded lengths to fit any band machine or in coils to be individually cut and welded as needed, Figure 17-3. The following terms are used to designate the five characteristics of a saw blade.

Tooth Forms

Regular or *precision tooth blades* are used for general-purpose metal cutting. *Buttress* or *skip tooth blades* have wide-spaced teeth to provide clearance for cutting softer materials including wood. *Hook* or *claw tooth teeth*

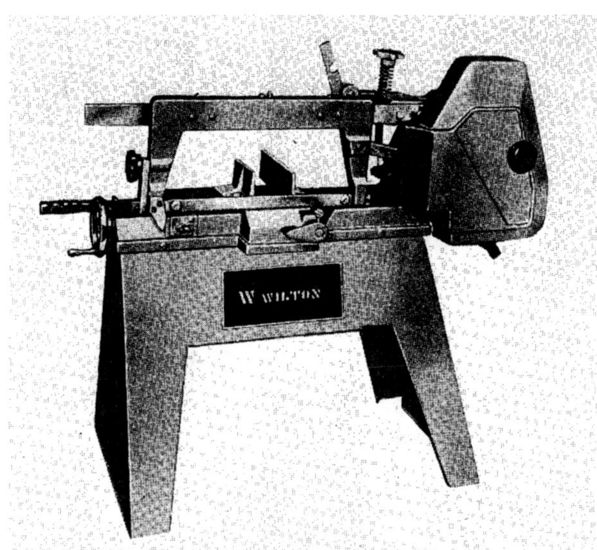

Fig. 17-1 Reciprocating power hacksaw (Wilton Corporation)

Fig. 17-2 Metal-cutting band saw (Wilton Corporation)

Fig. 17-3 Band saw blades (Do All Company)

have deeper, rounded gullets that tend to curl the chips. This tooth form is good for sawing harder, nonferrous alloys, hard woods, and plastics. *Combination* or *vari-pitch tooth* designs have varying set angles and gullet depths. Vibrations are broken up with this tooth design, resulting in smoother, quieter cutting and additional blade life, Figure 17-4.

Pitch

Pitch of the blade refers to the number of teeth per inch. Thin materials require a finer pitch or greater number of teeth per inch. Thick materials require a coarse pitch with relatively few teeth per inch, Figure 17-5.

Blade Width

The *width* of a blade, Figure 17-6, determines the ability of the blade to cut straight or contour. The narrower the blade width, the smaller the radius that can be cut. The wider the blade, the more accurate the cut is.

Blade Thickness

The *gage* or thickness of a saw blade varies with the blade width, Figure 17-6. Band blades 1/2 inch wide or less are usually 0.025-inch gage; 5/8-inch and 3/4-inch wide blades are generally 0.032-inch gage; 1-inch wide blades are 0.035-inch gage. The heavier the gage, the wider the width of the blade; therefore, the greater the beam strength.

Set

Set refers to the amount offset given the teeth to provide clearance for the back of the band. Refer to Figure 17-6. The three common sets are raker, wavy, and alternate. Refer to Figure 14-30.

The *raker set* is preferred for cutting solid cross sections that are uniform in shape. It is also used for contour or profile cutting operations. *Wavy set* gives greater chip clearance and helps eliminate tooth strippage. This set is recommended for most general-purpose horizontal band machines. They are best used in work where the cross section

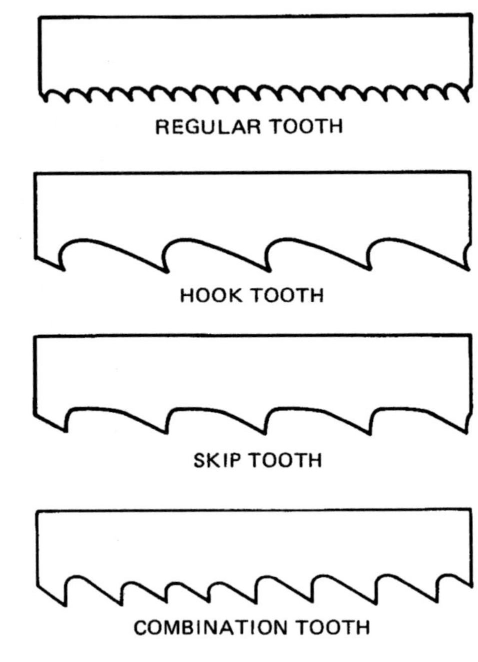

Fig. 17-4 Tooth forms

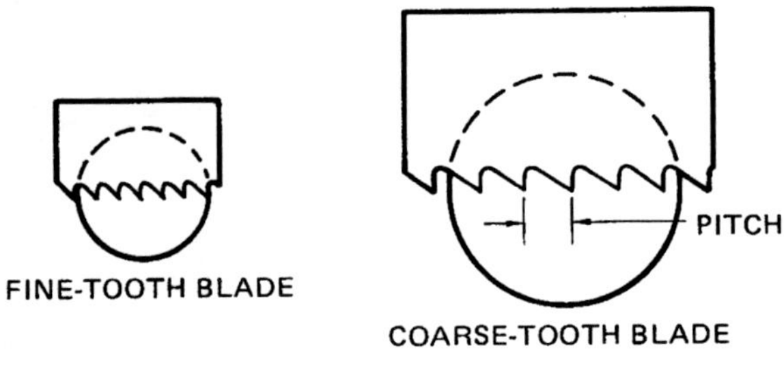

Fig. 17-5

Fig. 17-6 Measuring the width and thickness of a saw blade

changes, such as pipe and structurals. The *alternate set* is used mainly to cut light, nonferrous materials including thin sheet metal, tubing, and plastics.

BLADE SELECTION

Correct blade selection is determined by the type of material to be cut, thickness of the material, type of cut desired, and the radius of the material.

CUTTING FLUIDS

The following are reasons for using cutting fluids when band sawing.

- Cutting fluids remove heat from the cutting operation.

- The fluids reduce friction from the rubbing action of the chip against the cutting surface of the band saw tooth.

- The coolant tends to dirty the area between the tooth and the chip. This prevents the pressure welding of the workpiece material to the saw tooth, called a *built up edge*.

- Provides longer life to the saw blade.

- Cutting fluids are used to flush chips from the cutting area.

- Cutting fluids lubricate the saw band to prevent scoring and binding as it passes through the work and the saw guides.

- Slow-speed band saw machining can be done dry. Cast iron should be cut dry because the free carbon in it makes it easy to saw at slow speeds.

HORIZONTAL BAND SAW

Figure 17-7 shows work held square in the saw vise being cut with the horizontal band saw. Work can also be clamped at an angle to the saw blade to produce an angular cut on the workpiece. The size of a band saw machine is determined by the largest size square the saw can cut through.

Floor band saws are powered by electric motors through V pulleys, belts, and a gear reducer. The correct speed of the saw blade for a variety of materials is obtained by changing the V belt from one pulley to another. Many new style saws manufactured today are equipped with variable-speed control for the blade. Convenient speed charts are often found attached directly to the power saw. New-improved band saw blades are continually being developed.

It is good practice to follow the manufacturer's recommendations for blade pressure and cutting speeds. This information is

Fig. 17-7 Work clamped in a horizontal band saw (DoAll Company)

available in technical bulletins or from a manufacturer's representative.

Using the Horizontal Band Saw

The following may be used as a guide for operating the horizontal band saw.

1. Leave enough stock so the part may be finished to required size in later machining operations. Also, be sure to select the correct kind and size of material before cutting.

2. Mark the work to the desired length. Insert the work in the vise.

3. Support both ends of long material extending from the saw vise with suitable floor stands.

4. Always set the blade guides as close as possible to the workpiece. This provides maximum support to the blade and greater accuracy of the cut.

5. Release the lock and lower the upper saw frame and saw to the work. Set the work to the saw and tighten the vise jaws. Most horizontal band saw machines have either a mechanical or hydraulic lock to hold the upper saw frame in an up position.

6. Check the weight or feed pressure and start the cut. Some saws are equipped with a controlled lowering and feed mechanism. As the saw dulls, the feed rate must be increased.

7. Chips are a clue to correct the feed pressure. If there are:
   - powdered chips, increase the feed.
   - blue-color burned chips, reduce the feed.
   - free curling chips (not discolored), the feed is just right.

8. Adjust the cutting fluid to the point of cut, Figure 17-8.

9. When several pieces of the same length are to be cut, the bar stop may be set to the desired length. (Refer to Figure 17-7.) The stop may be swung clear when not in use.

10. Material may also be stacked in the vise to cut multiple parts, Figure 17-9.

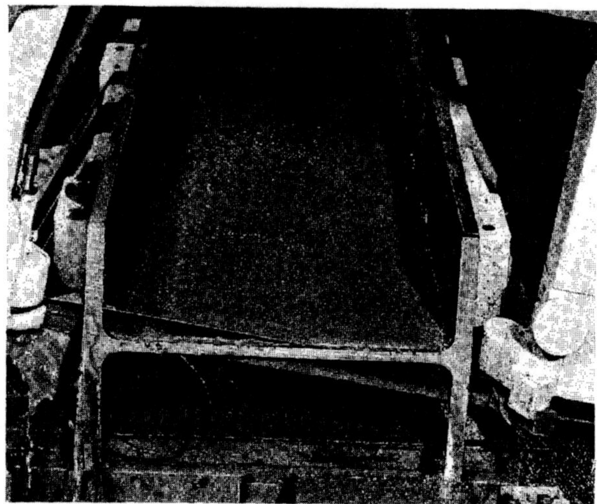

Fig. 17-8 Cutting fluid being applied at the point of cut (DoAll Company)

Fig. 17-9 Material stacked in the vise to cut multiple parts

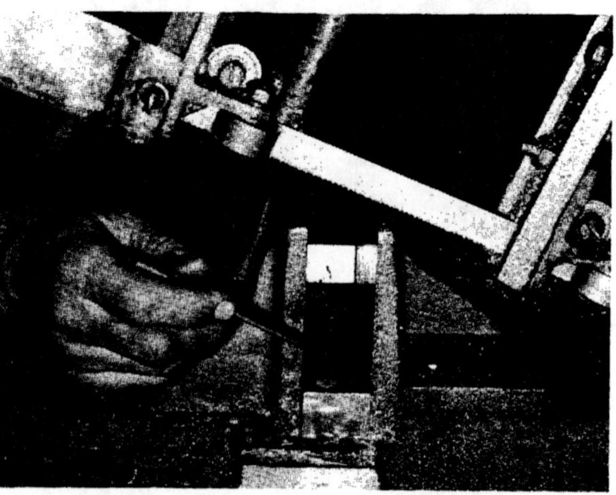

Fig. 17-10 Supporting the free ends of the vise with a piece of scrap when holding short pieces of material to be cut

11. Remove gloves and have the power off when making any band saw blade adjustments or changes.

12. Never start a new blade in an old cut. Rotate the material to a new location and start a new cut. The old cut is narrower than the set of the new blade and quickly dulls the new teeth.

13. Always support the free end of the vise with a piece of scrap when holding short pieces of material to be cut, Figure 17-10.

Blade Installation for the Horizontal Band Saw

1. Select the proper blade for the material to be cut.

2. Unplug the electrical supply to the machine or turn off the disconnect switch.

3. Loosen the blade tension adjustment and remove necessary guards. Remove the old blade using gloves.

4. Clean the band wheels and roller guides.

5. Install the new band blade making certain the teeth are in the cutting direction. If they are not, the blade must be twisted inside out.

6. Install the band over the two band wheels and twist the blade up into the roller guides.

7. Adjust the blade tension. Refer to the manufacturer's recommendations. Some saws are made with indicators, some count the turns of the adjusting wheel, and other manufacturers recommend tapping the blade with a hammer and listening for a definite ring.

8. Replace safety guards and connect the electrical supply.

9. Start the machine and check the blade for proper operation. Make necessary adjustments and begin sawing operations.

10. The new blade must be retightened after use. This adjustment takes up the stretch of the new band.

Also available for use is a hand-held, electric-powered, portable, metal-cutting band saw, Figure 17-11. This saw provides speed and convenience to many miscellaneous sawing operations where the material cannot easily be moved to the saw.

VERTICAL BAND SAW

The vertical band saw machine, Figure 17-12, is used for shape cutting operations. The horizontal band saw is generally used for cut-off work.

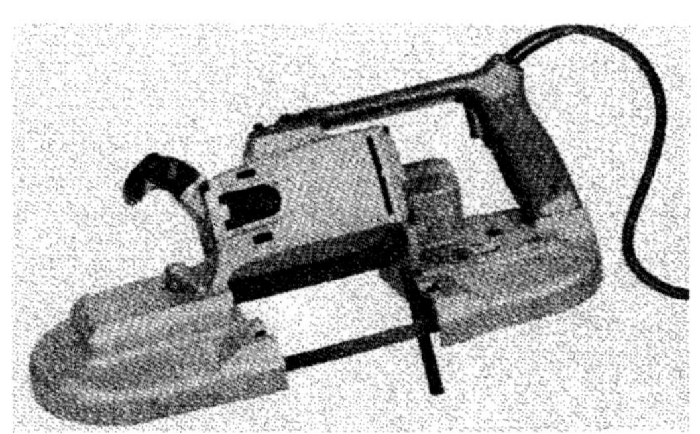

Fig. 17-11 Portable, metal-cutting band saw (Nicholson File Company)

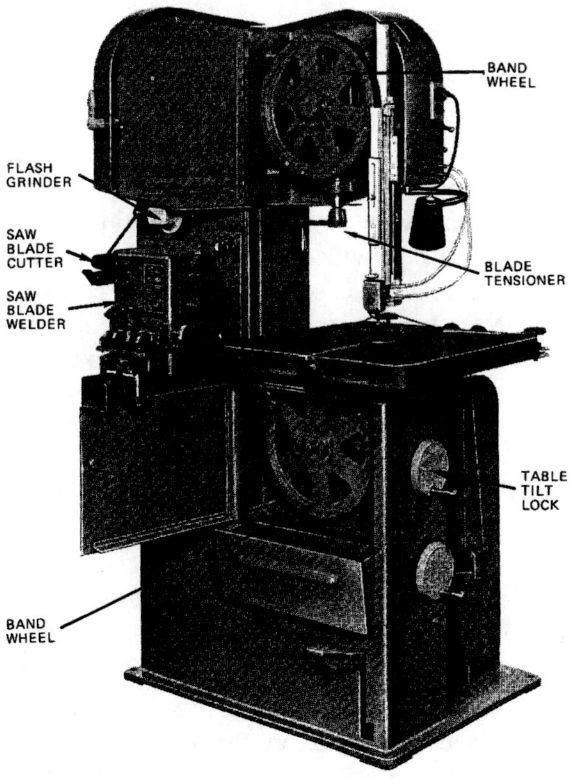

Fig. 17-12 Vertical band saw (DoAll Company)

Figure 17-13 shows a closeup of the sawing selector dial. This selector dial gives prompt information to the operator. When using the vertical band saw, work can be mounted on the table and fed square to the saw blade or mounted at an angle in a table vise and fed to the saw. Other types of cutting that can be done on the vertical band saw are contour cutting and slot cutting.

Blade Installation to Remove an Internal Workpiece Section

At times, the center section of a part must be removed without sawing through the outer section, Figure 17-14.

1. Drill a small hole in the center section large enough to accept the required saw blade.
2. Cut the blade and insert it through this hole and reweld at the machine.
3. Adjust the blade over the saw carrier wheels through the saw guides.
4. Tighten the blade by adjusting the tension handwheel to the correct tension shown on the indicator.
5. Machines without the band-tension indicators must be adjusted just enough to prevent the blade from twisting or wandering while cutting. Manual blade tension is provided by the handwheel and the tension is set with a graduated plate.
6. New blades stretch after use and need to be readjusted for correct tension. The blade must not be overstrained. The use of a torque wrench is recommended to assure uniform tension.
7. The blade is now ready to cut the internal shape from the part.

Blade Installation Procedure for General Vertical Sawing

1. Lift the cover on the saw exposing the band wheels.
2. Place the band on the wheels with the teeth pointing down toward the floor and pointing in the direction of band-wheel rotation.

Fig. 17-13 Closeup of a sawing selector dial

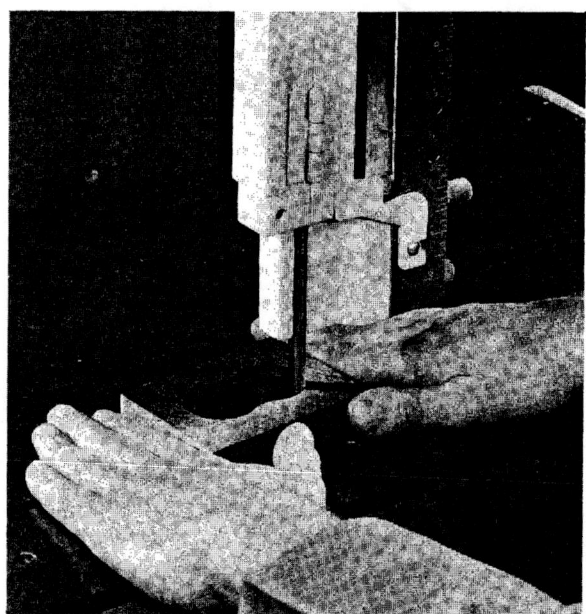

Fig. 17-14 Removing the center section of a part without sawing through the outer section (DoAll Company)

3. Twist the lower part of the band and insert it between the saw guides at the lower end of the saw-guide arms.
4. Turn the tension handwheel and adjust the blade to the required tension.
5. Adjust the saw guides in accordance with the manufacturer's manual.
6. Select the proper speed from the speed selector dial and adjust for that speed.

Vertical Band Sawing Procedure

1. Adjust the table to the cutting conditions — level or tilted to make an angular cut.
2. Use the correct blade and speed.
3. Always be sure the band-guide inserts are the correct size for the saw band used and they are properly adjusted.
4. Before starting the machine, adjust the upper band guide at a height to clear the work 1/8 inch (3.1 mm) to 3/8 inch (10.0 mm). The closer the guide to the work, the greater the accuracy.
5. Start the machine and feed the work to the saw gradually. Increase the feed slowly to the correct pressure after the saw has started the cut. Sudden changes in feed pressure may cause the band to break.
6. Lubricate the saw band and guides.
7. Use cutting lubricants and coolants as recommended by the machine manufacturer. Cutting fluid is determined by the band speed:

Band Speed	Type of Cutting Fluid
Less than 175 fpm (feet per minute)	Straight oil
150 – 300 fpm	Soluble oil
More than 250 fpm	Chemical fluid

Note: Additional information on cutting fluids may be obtained from suppliers.

Welding the Vertical Band Saw Blade

Nearly every vertical band machine made has a built-in butt welder. The term *butt welder* means butting the ends of a saw blade together and then welding them. The weld is made by resistance of electric current passing through the saw band. This resistance generates enough heat to weld the ends of the blades together.

With experience, a band can be cut and welded in about 4 minutes. The weld in the band must be as strong as the band itself. Some of the larger, heavy-duty welders are also known as *flash butt welders.*

The following steps may be used as a guide when welding the band blade:

1. Select the correct band material, tooth form, and pitch for the job.
2. Pull the desired length of band from the storage box.

> **Caution:** Use gloves to prevent cuts from the sharp saw teeth.

3. Mark the blade to the correct length and cut it off square using the built-in saw blade cut-off shear. A pair of snips may also be used to shear the blade to length.
4. Square the snipped ends by looping the cut length band to bring the two ends together with the teeth pointing in opposite directions. Grind both ends together, Figure 17-15.
5. Grind off a few of the teeth on each side of the cut before making the weld,

Fig. 17-15 Ends are ground together and then thinned to correct tooth fit (Do All Company)

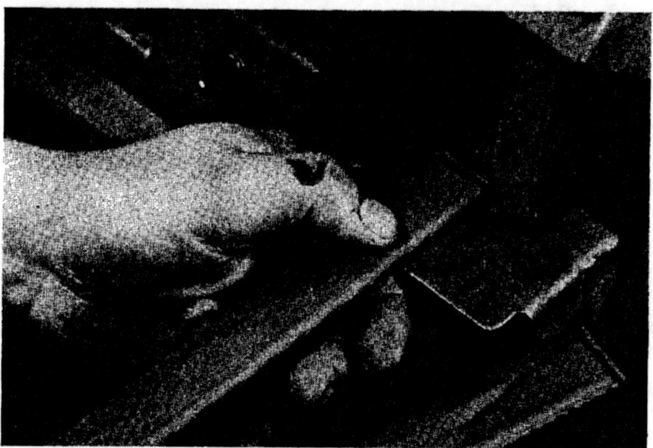

Fig. 17-15A Grinding off a few of the teeth on each side of the cut before making the weld (Do All Company)

Figure 17-15A. The weld consumes about 1/4 inch (6.3 mm) of material in making the weld. This loss of material must be considered to nsure uniform tooth spacing in the blade after the weld has been made.

6. Clean the ends of the blade free of any oil film. Oil prevents good electrical contact between the blade and the welder jaws and causes a poor weld.

7. Clean the jaws of the welder.

8. Set the welder to the correct settings for the blade to be welded. This information is found on the band welder.

9. Place one end of the blade in the right jaw with the teeth to the rear against the aligning surface at the back of the jaw. Place the end of the blade in the center of the gap between the two jaws. Clamp this blade in the jaw. Repeat this procedure with the other end of the blade placed in the left jaw.

10. Make sure the band is not twisted as the left end is inserted into the welder. Also, make sure the blade teeth are tight against the aligning surfaces. This insures the band will be straight after welding.

11. Check the position of the blade ends. They should be just touching in the center of the gap between the jaws. The ends must not be offset or overlapped.

12. Press the weld switch all the way to a full mechanical stop. Release it immediately.

> **Caution:** Step aside to avoid the weld flash. Be sure to wear safety glasses and a face shield.

13. The weld is automatically made and the blade has become a band.

14. The weld flash should be consistent with no sputtering. The flash should be sharp and bright, not dull in color.

15. Remove the band from the welder. Inspect the weld as follows:

- Make sure the upset material from the weld across the band is uniform, Figure 17-16.

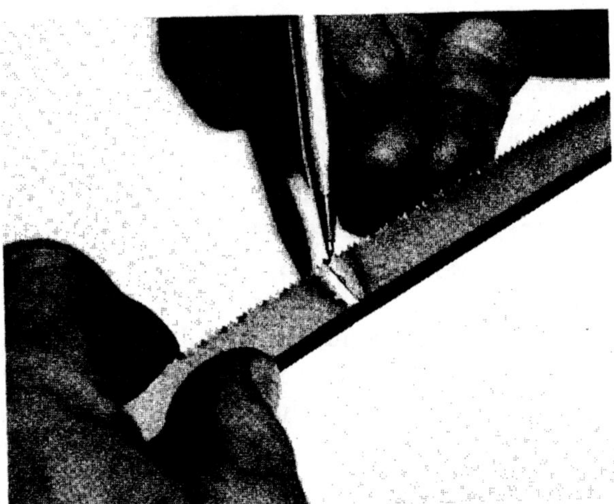

Fig. 17-16 Note the uniform upset of the weld across the band

- Make sure the color of the upset material is the same blue gray throughout.

- Be certain that the spacing of the teeth is uniform and the weld is located in the center of the gullet.

- Make sure the band is straight. Check this with a straightedge. If it is not, the weld is no good. Cut out the weld area and start over.

Annealing Procedure

The band must be annealed and the welding flash ground away before it can be used. Welding leaves the blade very hard and brittle at the weld.

Annealing is a process of reheating the blade and slowly cooling it to relieve stresses and eliminate hot spots which cause the blade to be brittle. Annealing softens the blade to a desired degree of hardness or toughness. The annealing temperature is detected by the color of the metal. Both carbon and high-speed steel blades require an annealing color of dull cherry red.

1. Clean the welder jaws and inserts.

2. Clamp the band in the jaws with the teeth to the rear and the weld centered in the gap between the two jaws.

3. Set the selector switch for the correct annealing heat.

4. Push in and jog the anneal switch button occasionally until the saw band reaches the correct color temperature. This takes about fifteen seconds.

5. Shield the weld area with your hand to be able to see the dull red color. Avoid overheating as the band will harden and become brittle.

6. Scale, oil, or grease on the weld may need to be removed in order to see the correct annealing color.

7. After the band becomes a dull cherry red release the switch button. As the weld area starts to cool, press in the anneal switch button from time to time to slow down the cooling process. Rapid cooling may cause the weld to become hard and brittle.

8. Gently grind away the weld flash using the built-in grinding wheel for this purpose. When grinding, hold the band with the teeth facing the operator. Avoid contact between the teeth and the grinding wheel.

9. Remove all the excess weld flash down to the original thickness of the band. Check the thickness with the thickness gage built in above the flash welder. Do not grind too much off and weaken the blade.

10. The band may now be installed on the machine, ready for the cutting operation. Discarded blades should be coiled and tied before placing them in scrap containers.

Special Operations

Other special operations done on the vertical band machine include polishing, filing, friction sawing, high-speed sawing, slicing with knife-edge blades, spiral-edge band sawing, line grinding, cutting with diamond-edge saw bands, and electro band machining.

Safety Precautions

The following safety precautions apply to all power sawing operations:

- Only those persons authorized and fully qualified should operate power saws.

- Safety goggles or a face shield must be worn when operating the power saw.

- When operating any machine, make sure you remove all jewelry before starting.

- Never readjust the saw or the work while the saw is in operation. Keep hands away from the saw blade while the saw is in operation, Figure 17-17.

- Always get help to move or place heavy stock to or from the saw.

- Always support the ends of long material with suitable supports.

- Do not clean the saw with bare hands.

- Remove sharp burrs from sawed stock with a file or grinder.

- Replace the band saw blade when it becomes dull, pinched, or burned.

- Keep loose clothing away from the moving blade.

- Keep all safety guards in place.

Fig. 17-17 Use pushing boards to move the work close to the saw blades. Keep fingers away from the saw blade.

REVIEW QUESTIONS

A. Multiple Choice

1. The blade guides should be set to the workpiece in which of the following ways?
   a. As close as possible
   b. Not important
   c. High
   d. Low

2. As a band saw blade dulls, the feed should be:
   a. increased.
   b. decreased.
   c. left the same.
   d. reduced by one-half.

3. Parts of the same length may be cut from the bar without measuring each part by using:
   a. the front guides.
   b. the adjustable stop.
   c. the rear guides.
   d. the feed control.

4. The recommended distance above the work to set the upper band guides when vertical band sawing is:
   a. 1/2 inch (12.7 mm) to 3/4 inch (19.0 mm).
   b. 3/8 inch (10.0 mm) to 5/8 inch (15.9 mm).
   c. 3/4 inch (19.0 mm) to 1 1/4 inches (31.7 mm).
   d. 1/8 inch (3.1 mm) to 3/8 inch (10.0 mm).

B. Short Answer

5. State the reason why the band saw blade must be clean from oil before butt welding.

6. After the blade is welded, state the two additional steps that must be done before using the blade.

7. State the four main advantages of the metal-cutting band saw.

8. State the two methods of changing speed on the power saws.

9. Give one reason for using a cutting fluid when band sawing.

ACTIVITIES

1. With your instructor, complete the following operation on the vertical band saw:
   a. Select the correct band saw blade for the job.
   b. Cut and weld the blade.
   c. Install the blade.
   d. Lay out and cut an assigned external part to shape.
   e. Prepare and cut out an internal section to size.
   f. Clean the saw and replace the unused materials.

2. With your instructor, complete the following operation on the horizontal band saw:
   a. Select the correct band saw blade.
   b. Install the blade on the saw.
   c. Set the work in the vise.
   d. Set the saw guides properly.
   e. Set the material stop.
   f. Cut the part to length setting the required blade pressure.
   g. Clean the saw and replace unused materials.

UNIT 18 BENCH AND PEDESTAL GRINDERS

OBJECTIVES

After completing this unit, the student will be able to

- identify the common grinders used in benchwork.
- demonstrate the correct and safe use of these grinders.
- explain the standard, grinding-wheel marking system.

INTRODUCTION

Bench and *pedestal grinders* (or floor grinders), Figures 18-1 and 18-2, are used for many grinding jobs at the bench. These jobs include sharpening tool bits, drill bits, chisels, screwdrivers, and shaping parts by removing unwanted material from a workpiece.

Bench and pedestal grinders are made the same except for their size and method of mounting. Bench and pedestal grinders are sized by the size wheel they are designed to hold. A 10-inch grinder uses a 10-inch wheel. These grinders are usually fitted with a roughing and a finishing grain abrasive wheel. Wire brushes, polishing, and buffing wheels can be substituted for the removable grinding wheels, Figure 18-3.

GRINDING WHEEL

A *grinding wheel* is a cutting tool. Each abrasive grain or crystal that makes up the wheel is a multi-edge cutting tool. As the wheel revolves, many cutting edges cut away chips from the workpiece. The size of the abrasive grains determines the coarseness or fineness of the wheel. The abrasives are held together in the required wheel shape by various bonding materials.

A standard marking system is used on all grinding wheels. The wheel marking is usually printed on the wheel blotter attached to the side of the wheel, Figure 18-4.

Example: A typical wheel marking may read as follows:

	A	46	–H	8	VBE	8
Position	1	2	3	4	5	6

This system contains the above six identifications of the grinding wheel and represents the following:

Position	1 (A)	— Abrasive type
	2 (46)	— Grain size
	3 (H)	— Grade
	4 (8)	— Structure
	5 (VBE)	— Bond
	6 (8)	— Manufacturer's record

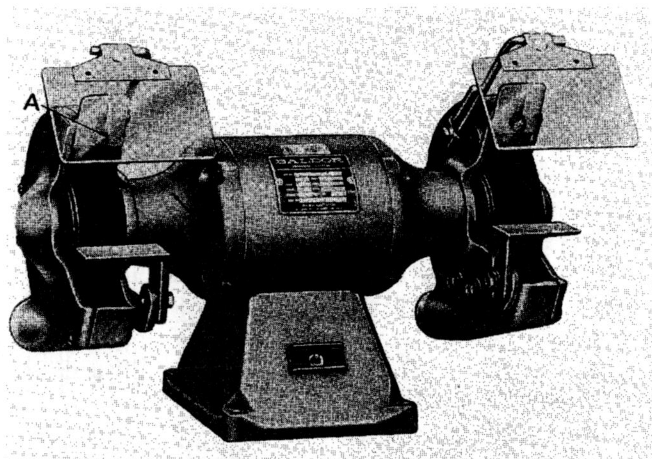

Fig. 18-1 Parts of the bench-mounted grinder: (A) Spark Arrestor, (B) tool rest, and (C) eye shield (Baldor Electric Company)

Fig. 18-2 Pedestal-mounted grinder (Baldor Electric Company)

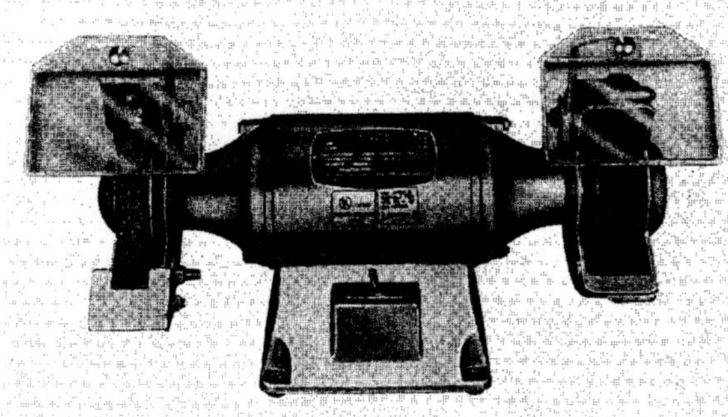

Fig. 18-3 Bench grinder with: (A) Polishing wheel, and (B) wire-brush wheel (Sioux Tools, Inc.)

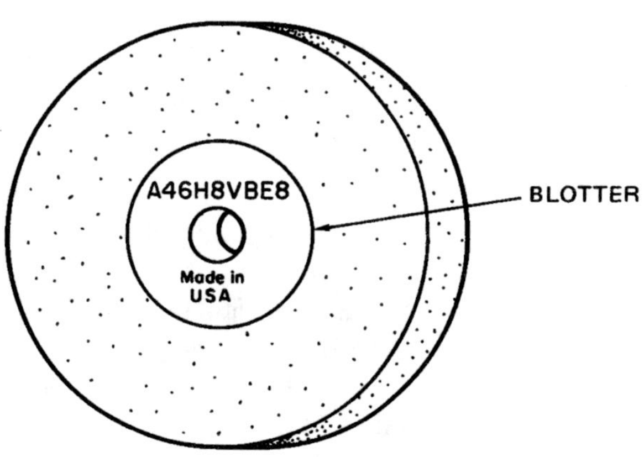

Fig. 18-4 Wheel marking printed on wheel blotter

Basic Abrasive Types

Aluminum oxide, represented by the letter A, is recommended for general-purpose grinding of high tensile-strength materials including steel and steel alloys. *Silicon carbide,* represented by the letter C, is used to grind low tensile-strength materials such as carbide and other hard, brittle metals including cast iron, bronze, and aluminum.

Grain Size

The size of the grain is named according to the size of the screen opening through which they are sorted. The 46-grain size would measure about 1/46 of an inch across. The screen would have 46 openings per linear inch, Figure 18-5.

Grain sizes are as follows:

- 10 through 25 — coarse
- 30 through 60 — medium
- 70 through 180 — fine
- 220 through 600 — very fine

Use fine grain size on hard materials and coarse grain size on soft, easily formed materials.

Grade

The *grade* is a rating of the strength of the bond which holds the abrasive grains in place. The rating is as follows:

- A - H = Soft
- I - P = Medium
- Q - Z = Hard

Use soft wheels on hard materials and hard wheels on soft materials.

Structure

The *structure* of a grinding wheel is defined as either dense or open. A *dense wheel* has the abrasive grains very tightly packed and appears to be smooth. An *open-grain wheel* has larger holes between the grains. The range of density is from 1 to 16, with 1 being the densest, and 16 with the most open spacing between the grains. Use close-grain structure wheels on hard, brittle material. Use open-grain structure on soft, easily formed materials.

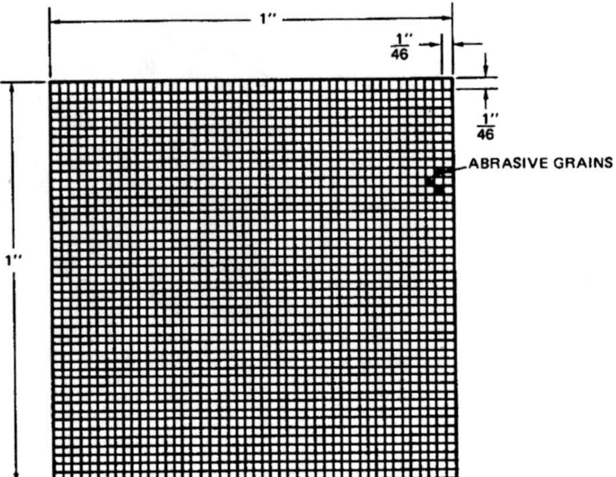

Fig. 18-5 46-grain size

Bond Type

The *bond type* refers to the type of material used to hold each grain to one another in the shape of a wheel. The five main bond types are:

- V (Vitrified) — Most commonly used for high stock removal

- S (Silicate) — Releases dull abrasives quickly, good for tool grinding

- B (Resinoid) — Cool cutting, high stock removal, can be run at high speeds

- R (Rubber) — Used for good finishes

- E (Shellac) — Produces high finishes on hard surfaces

Depending upon the type of bond, it is the amount of bond material and combination of bond materials that determines the hardness or softness of a grinding wheel. Wheels from which the abrasive grains are easily broken away are known as *soft grade.* Those wheels which retain the grains are called *hard grade.*

Manufacturer's Record Number

The *manufacturer's record number* is optional and not always shown on the wheel marking. With this number, the manufacturer can quickly refer to when and where the wheel was made.

Procedure to Mount the Grinding Wheel

1. Select the correct wheel for the type of grinding required. Refer to wheel manufacturer's current recommendations.
2. Visually inspect the wheel for damage or cracks.
3. Ring test the wheel for hidden cracks. This test is done by suspending the wheel from the hole on a small pin or the finger. The wheel is tapped gently with a wooden screwdriver handle. Heavier wheels are allowed to rest in a vertical position on a clean, hard floor, Figure 18-6. Use a wooden mallet for heavy wheels. The best position to tap the wheel is about 45 degrees either side of the vertical centerline. Rotate the wheel 45 degrees and repeat the test. A sound and undamaged wheel gives a clear, metallic tone or ring. A cracked wheel gives off a dull, thud sound.
4. Clean the shaft and wheel flanges. The wheel flanges must be equal and at least 1/3 diameter of the wheel and relieved around the hole, Figure 18-7. The inner flange must be securely fastened to the shaft.
5. Use the mounting blotters supplied with the wheel.
6. Blotting paper washers must not be thicker than 0.025 inches (0.6 mm). The blotter is used to insure even pressure on the wheel and dampen vibration between the wheel and the shaft.
7. Mount the wheel with the blotter on the shaft. Never force the wheel on the shaft. A clearance of 0.002 inch (0.05 mm) to 0.005 inch (0.12 mm) should be provided.
8. Install the outer wheel flange, being sure the outer blotter is in place on the wheel.
9. Install the washer and securing nut.
10. Tighten the securing nut only tight enough to hold the wheel firmly. Overtightening may damage the wheel.

Note: The thread on the right end of the grinder shaft is right-handed. The thread on the left end of the shaft is left-handed. This prevents the locking nuts from loosening during operation.

11. Replace the safety guards. Adjust the spark arrestor and tool rest to 1/16-inch (1.5 mm) wheel clearance. Make certain the tool rest is set on the horizontal centerline of the wheel. (See Figure 18-1.)
12. Stand to the side of the wheel while turning on the grinder.
13. Allow the newly mounted wheel to run at operating speed for at least 1 full minute before grinding. A cracked wheel is most likely to explode or fly apart during this time.

Dressing the Wheel

The newly mounted wheel must now be balanced or *trued*. The truing operation

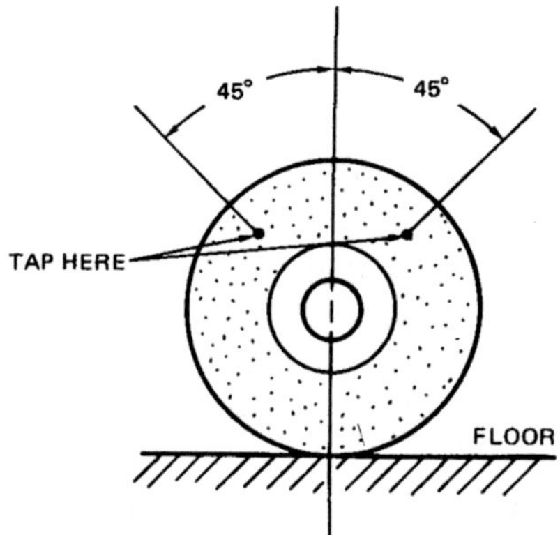

Fig. 18-6 Large, heavy wheels are supported on a clean, hard floor.

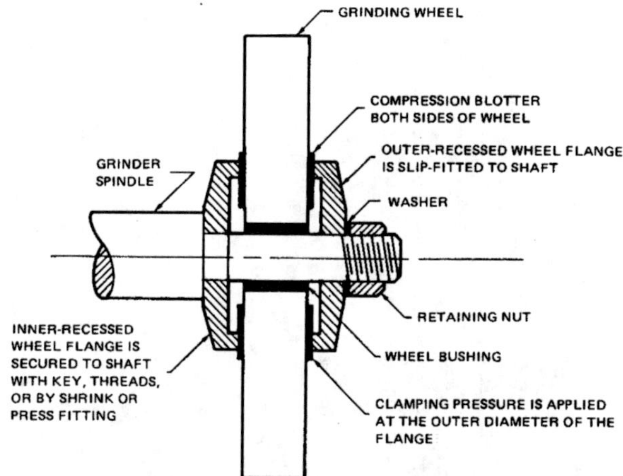

Fig. 18-7 Correctly mounted grinding wheel with blotters and wheel flanges in place

removes material from the high side of the wheel. This removes any material that is not round and causes the wheel to run true.

As the wheel is used, the cutting abrasives dull, and the wheel becomes glazed. The wheel may also become loaded with grinding chips that wedge between the cutting grains. This condition occurs in the grinding of brass, bronze, aluminum, other soft materials, and some steels.

An ideal wheel would automatically sharpen itself through grinding pressure. The

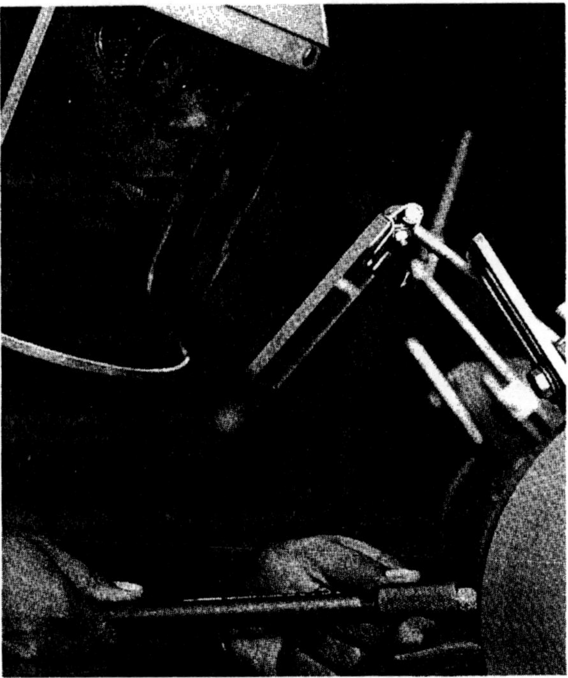

Fig. 18-8A Abrasive stick used to finish dress the face of the grinding wheel

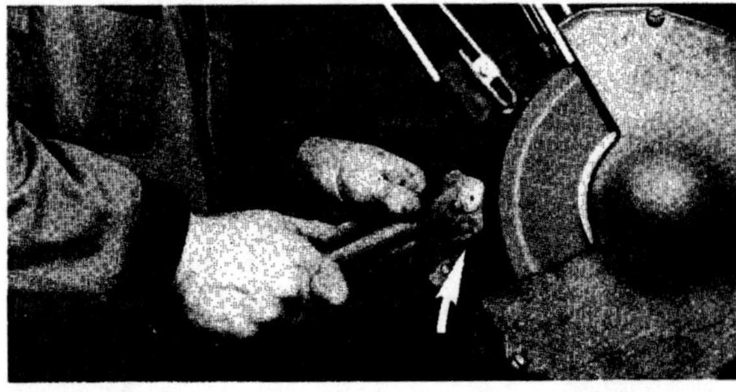

Fig. 18-8 Correct use of the star wheel dressers. Note the arrow indicating the tool rest that guides the dresser.

dull abrasives would be broken out and new sharp crystals exposed. Removal of these dull abrasive grains and wedged grinding chips in the wheel can be done by dressing. Many types and sizes of wheel dressers are used in the shop.

The *mechanical dresser*, Figure 18-8, uses the action of revolving pointed steel discs or star cutters mounted on an arbor and separated by thin washers. This tool is pressed to the rotating grinding wheel and moved back and forth to clean the wheel face. The tool rest is used as a guide for the dresser. Be sure to use enough pressure against the wheel so as not to just grind away the steel cutter discs of the dresser. *Diamond dressers* and *abrasive stick dressers* are also used to finish dress grinding wheels, Figure 18-8A. They are used in a finishing manner similar to the mechanical dressers.

After dressing the grinding wheel, retighten the arbor nuts. Readjust the tool-rest support plate and spark arrestor to keep the wheel face within 1/16-inch (1.5 mm) to 1/8-inch (3.1 mm) clearance. Also, reset the grinding shields.

Correct Use of the Grinding Wheel

1. Grind, using the full face of the wheel. Avoid making ruts or grooves into the wheel.
2. Grind only on the cutting face of the wheel, not on the sides of the wheel.
3. Always grind away from the cutting edge of a tool.
4. Never jam a workpiece into the wheel.
5. Keep wheels in good condition.
6. Choose the right wheel rated at higher speeds than it will be running in operation.
7. Use a grinding wheel only for the material for which it was designed to grind.
8. Keep all safety guards in place.
9. Always handle and store grinding wheels carefully.
10. Be sure the grinder is correctly grounded to prevent electrical shock.

Procedure to Sharpen Prick and Center Punches

Grinding procedures to sharpen a prick punch, center punch, and screwdriver point are as follows:

1. The *prick and center punches* are ground with cone-shaped points. Adjust the grinder rest so that the end of the punch is ground to the desired angle.
2. The punch is placed on the tool rest and rotated as it is pressed against the grinding wheel, Figure 18-9.
3. The *screwdriver blade* is ground by first squaring the end.
4. Next, adjust the tool rest to an angle to produce parallel sides on the point of the screwdriver blade. This allows the point to correctly fit into a screw head slot.

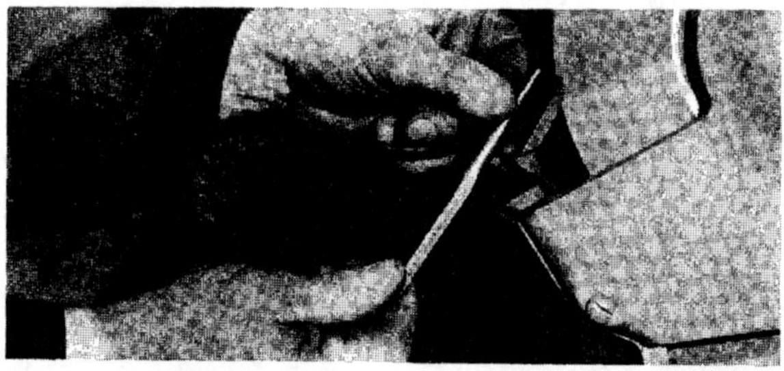

Fig. 18-9 Correct way to grind the prick punch and center punch

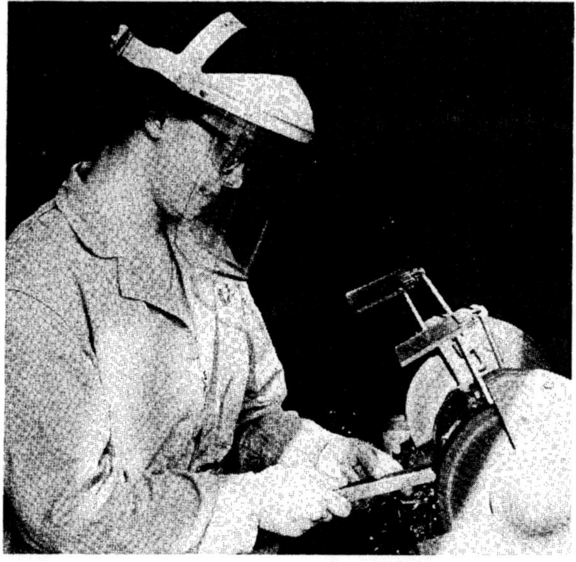

Fig. 18-10 Properly dressed for grinding operation

Safety Precautions

- Grinding dust is hazardous to health and damages precision parts. When available on a grinder, use the grinding dust collector.

- When grinding tool points, always take care not to overheat the point. Dip the point in cold water frequently to prevent heat build up.

- Always adjust the grinder correctly and wear proper eye protection and clothing, Figure 18-10.

OTHER TYPES OF GRINDERS

The *surface grinder* is occasionally used by the benchworker to grind surfaces flat, Figure 18-11. The work is held on a magnetic table and moved back and forth under the revolving grinding wheel. Coolant is supplied to the work and wheel through a pump system. The wheel is very accurately fed to the work by an adjusting handwheel.

When surface grinding, remove all burrs from the work and the magnetic chuck to insure accurate seating of the work. Make sure the magnetic chuck is turned on by trying to remove the work. Be certain that the workpiece lays across at least 3 magnetic bars of the magnetic chuck for each 1 inch (25.4 mm) of workpiece height.

Fig. 18-11 Surface grinder

Note: A piece of paper between the workpiece and the magnetic chuck serves to hold the work more securely. Before using the surface grinder, consult with the manufacturer's instruction manual for the correct and safe operation.

Figure 18-12 shows an *abrasive cutoff saw*. This saw uses an abrasive grinding wheel to cut materials into required lengths.

The cold saw uses a circular metal blade rather than an abrasive wheel. The blade cuts a narrow kerf which saves expensive materials.

The *cold saw* is similar in design to the abrasive saw in Figure 18-12. The cold saw is made with a heavier, more rigid frame to make cuts with greater accuracy.

Fig. 18-12   Abrasive cutoff saw (Wilton Corporation)

REVIEW QUESTIONS

A.   Multiple Choice

1. The tool rest may be positioned in which of the following ways?
   a. Moved in
   b. Moved out
   c. Tilted
   d. All of these

2. Which of the following is the correct vertical height setting of the tool rest with regard to the horizontal centerline of the wheel?
   a. Makes no difference
   b. Above the horizontal centerline
   c. The same as the horizontal centerline
   d. Below the horizontal centerline

3. The spark arrestor and the tool rest on the grinder should be set at
   __________ from the grinding wheel.
   a. 1/8 inch (3.1 mm) to 3/16 inch (4.8mm)
   b. 1/16 inch (1.5 mm) to 1/8 inch (3.1 mm)
   c. 3/16 inch (4.8 mm) to 5/16 inch (8.0 mm)
   d. 1/8 inch (3.1 mm) to 1/4 inch (6.3 mm)

4. Before mounting a wheel, it must be checked for:
   a. color.   c. weight.
   b. cracks.   d. shipping numbers.

5. What must be done to a grinding wheel that has become loaded?
   a. Shaped   c. Dressed
   b. Glazed   d. Trued

B.   Short Answer

6. State the two differences between the bench and pedestal grinders.

7. State the reason for a left-hand thread on the left end and a right-hand
   thread on the right end of the grinder shaft.

8. What are the two basic types of abrasive grains used to make grinding
   wheels?

9. State the two reasons why it is important to use blotters when mounting
   a grinding wheel.

ACTIVITY

1. From four different grinding wheels selected by your instructor, identify the following:
   a. Abrasive type   d. Bond
   b. Abrasive grain size   e. Manufacturer's record
   c. Structure

UNIT 19 SPECIAL BENCHWORK TOOLS AND PROCEDURES

OBJECTIVE

After completing this unit, the student will be able to

- identify special tools and procedures used in benchwork.

INTRODUCTION

There are many safe, improvised short-cuts used to get work done at the bench. This unit is a summary and conclusion to present a few of the special tools and procedures used in benchwork.

The use of improvised methods of completing a given task are many times necessary. They are best learned through experience and everyday practice in the trade. Confidence is gained through this practice and experience.

PROBLEM-SOLVING APPLICATIONS

It would be nearly impossible to list the solution for every trade problem. What the worker does need to learn is how to solve everyday shop problems. This skill of problem solving requires a positive attitude towards logical solutions. This attitude is needed to complete most any work. The rules of common sense must also be applied in any problem-solving task. A positive attitude towards yourself, your job, and your fellow workers also helps to get the job done more effectively in any trade.

The successful worker must also use *visualization*. This is the ability to actually see a job already completed before starting. This aids the worker in forming a workable procedure to complete the job to the required specifications. The benchworker must also reason out problems by reducing them to their very lowest terms.

For example, a large workpiece or machine may appear to be complicated until you begin to look at it as many smaller pieces and parts. Each small part becomes a much smaller problem to solve. Keep trying to solve the problem at hand.

COMMON PROCEDURES TO AID THE BENCHWORKER

1. Use every available chart and table. These provide a quick, easy and accurate method to get information.
2. Always contact suppliers or manufacturers to find quick answers to machinery problems.
3. Always keep hand benchwork to a minimum. Labor is expensive. Machines

can usually complete more work quickly, safely, and accurately than can be done by hand.

4. Learn to use an oxyacetylene torch safely and properly. Many parts can be assembled and disassembled easier at the bench with heat properly applied from the torch.

5. Basic electric welding should be learned. Many times a worn or undersize part needs to be welded and machined. Also, broken parts may require welding.

6. Learn a working knowledge of practical metallurgy, heat treating, and metal-spray buildup.

7. Learn to safely use the jacks, hoists, chain blocks, and wedges in your shop. You may be required to use these tools on heavy-duty jobs.

8. Use the shop vacuum cleaner for cleaning up small, sharp chips.

SPECIAL TOOLS

The *flexible extension gripping tool* is a small tool used to aid the benchworker, Figure 19-1. Common tweezers also make a handy tool to hold and assemble small parts. *Soldering irons* are useful in some types of work to join one part to another.

Rethreading dies are made in both metric and American standard sizes, Figure 19-2. These dies are used to clean external threaded parts free of nicks and burrs. The thread file in Figure 19-3 is also used to remove nicks and burrs from external threaded parts. *Rethreading taps* are used to clean out damaged or dirty internal threads, Figure 19-4.

SPECIAL PROCESSES

Methods of Removing a Broken Tap

1. If the tap has only started into the hole and breaks, it is usually removed with a punch and hammer to break it up or reverse the short, broken pieces out of the hole. If the hole is a through hole, the broken tap may be driven out by

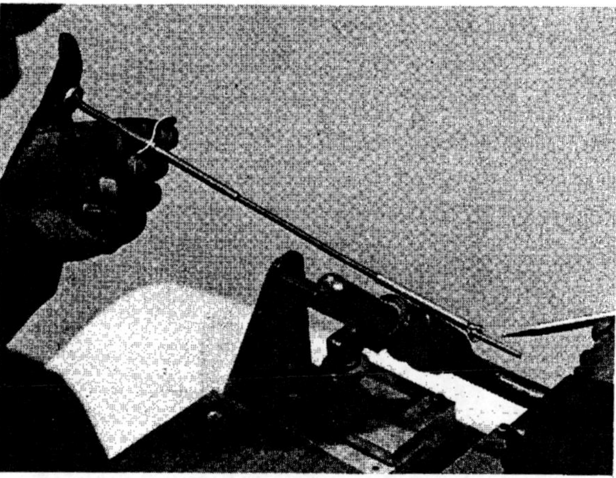

Fig. 19-1 Flexible extension gripping tool used to insert a bolt into a hard-to-reach part

Fig. 19-2 Standard set of rethreading dies
(Jaw Manufacturing Company)

Fig. 19-3 Thread file with plastic handle
(Jaw Manufacturing Company)

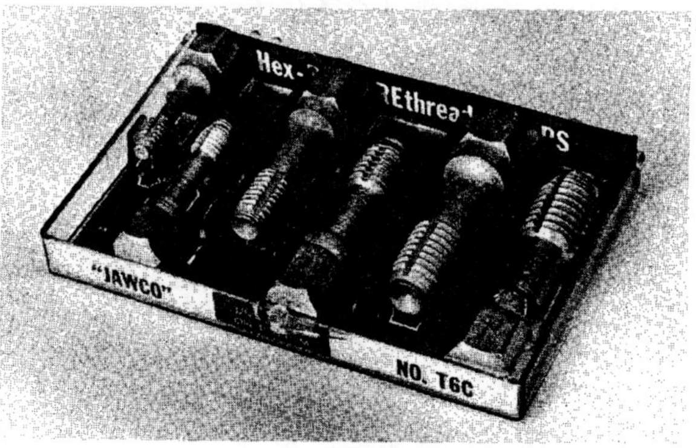

Fig. 19-4 Rethreading taps
(Jaw Manufacturing Company)

striking the end of the broken tap with a punch through the hole with a solid blow from a hammer, Figure 19-5.

2. After thoroughly cleaning out chips and broken tap particles with air, a broken tap can be removed by carefully reversing

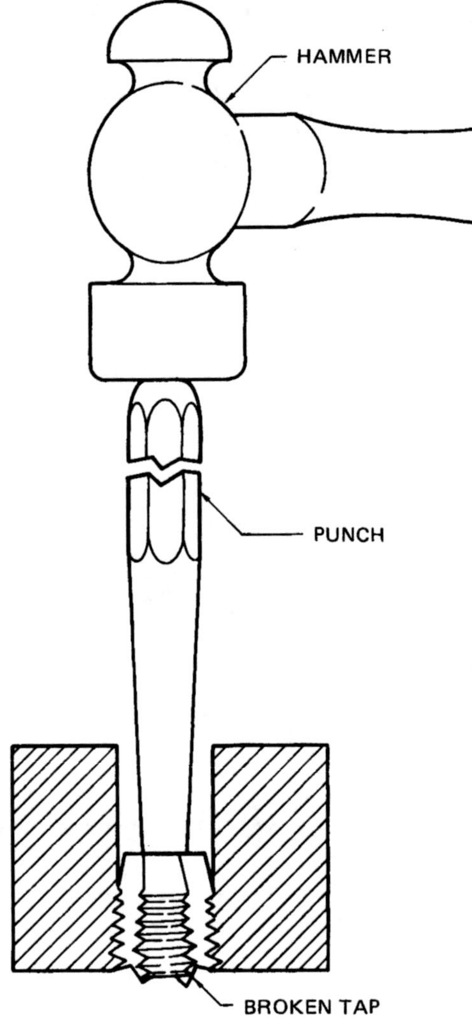

Fig. 19-5 Removal of a broken tap

the tap with a punch and hammer, Figure 19-6.

3. The broken tap may also be removed with the use of vise grips locked to the top of the broken tap. Enough of the broken tap must extend above the hole to provide a good grip, Figure 19-7.

Note: Commercial freeing-agent chemicals can be applied to help loosen the tap.

4. Another method is to add, by electric arc welding, additional metal to the shank of the broken tap above the level of the hole. Build up enough metal to weld on a nut and then proceed to back the tap out with a wrench. Make sure weld metal is not deposited on the threads in the tapped hole. This can only be done with carbon taps.

5. A *tap extractor* may also be used if the tap is broken below the top surface of the hole, Figure 19-8. First, clean out any chipped or broken particles of the remaining tap in the hole. Use the convenient magnet on the end of your scriber and air pressure. Loosen any other lodged parts or chips with a drift punch.

> **Caution:** Be sure to wear safety goggles and a face shield for any tap removal procedure.

After all the chips have been removed, place the tap extractor fingers into the flute of the broken tap as deeply as possible. Lubricate the broken tap thoroughly with penetrating oil. Place the collar of the extractor against the work surface. Twist the extractor back and forth carefully to try to loosen the tap in the hole before backing the tap out.

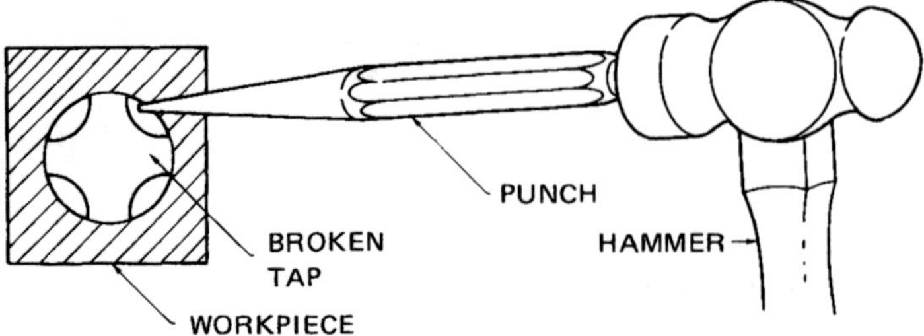

Fig. 19-6 Removing a broken tap with a punch to reverse the tap from the hole

Fig. 19-7 Vise grips are used to remove broken taps, stud bolts, and screws.

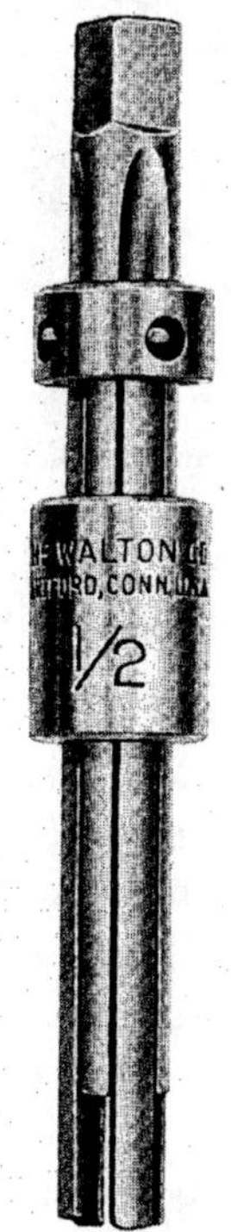

Fig. 19-8 Tap extractor (Walton Company)

The broken tap is then backed out by turning the extractor with a tap wrench. The tap extractor should not be forced. Otherwise, damage to the tap extractor fingers results.

If this method does not move the tap, it will have to be removed by one of the three methods described below.

1. If the tap material is carbon steel, the tap may be heated to a cherry red color with the oxyacetylene torch and allowed to slowly cool. This anneals or softens the tap so that it can be drilled out. Be sure to grind the top of the tap flat. Next, center punch and center drill the tap so the drill bit starts and follows straight through the center of the broken tap. Once the hole is drilled (less than the minor diameter of the tap), the remainder of the tap can then be collapsed with a punch and hammer or possibly removed with a screw extractor.

2. If the tap is made of high-speed steel, it may be removed by causing it to break up into small parts. This is done by heating the tap to a bright cherry red with the oxyacetylene torch and quickly quenching it in cold water. This rapid cooling of the tap causes it to become very hard and brittle. The tap may break up into small pieces and be easily removed. Be careful when heating the tap in this operation, however, as excessive heat may damage the workpiece.

> **Caution:** Always wear eye and face protection to prevent severe personal injury.

3. Probably the most effective method of broken tap removal is with the EDM (Electro Discharge Machine). In this process, metal is removed by electric spark erosion. The broken tap is eroded away by a round-shaped carbon electrode. The spark erodes or cuts a hole through the tap in a few minutes. These machines are also being used to produce many difficult shapes in metal parts.

Removing a Broken Stud

A broken stud may be removed with the use of the screw extractors, Figure 19-9.

1. Level the broken stud end by grinding or filing the end flat.
2. Center punch the end of the stud. Center drill the end of the stud.
3. Drill the proper hole size in the stud. The correct size is stamped on the screw extractor.
4. Install the screw extractor using a hammer to lightly enter it into the drilled hole.
5. With the use of a tap wrench or an adjustable wrench, turn the extractor in a counterclockwise direction. This procedure is used for removing broken right-hand threaded studs or bolts. For left-hand bolts, reverse the turning motion and use left-hand screw extractors. If the correct-size screw extractor is not available, a square, lathe-tool bit can be used. Grind the bit as shown in Figure 19-10.

The broken stud may also be removed by heating it to a cherry red with the oxyacetylene torch and allowing it to cool. The contraction of the stud is greater than the expansion. When cooled, it shrinks and can often be removed with the fingers.

Studs are frozen or stuck tight into a part may be removed by using two nuts jammed together, Figure 19-11. Broken off pipe plugs may also be removed with this shrink method. Broken pipe fittings are heated on the inside. After cooling, they are easily removed.

A nut can sometimes be welded to the top of the broken stud and the stud removed with a suitable wrench.

Other Removal Procedures

To remove a rusted or frozen nut from a bolt, saw off part of the nut, Figure 19-12. Be careful not to cut the threads of the bolt. Sometimes a nut that is stuck can be removed by tightening the nut slightly with a wrench

Fig. 19-9 A screw extractor (Greenfield Tap & Die, Division of TRW, Inc.)

Fig. 19-10 Square, lathe-tool bit

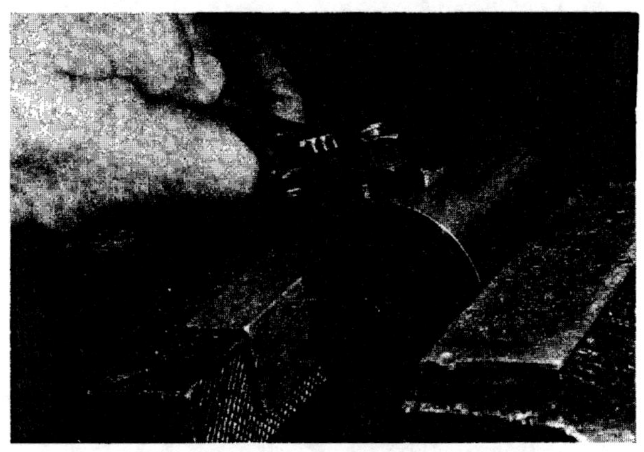

Fig. 19-11 The stud is removed by turning the lower nut.

Fig. 19-12 Removal of a rusted-on nut

and then loosening the nut. The same is true for a cap screw threaded into a part.

MISCELLANEOUS BENCH PROCESSES

Figure 19-13 shows the procedure to remove the center section of a workpiece. First drill to a prescribed layout line. Then cut between the drilled holes with a chisel. Finish this procedure by filing.

A *gasket* is a material placed between two parts to seal against leakage of a gas or liquid. Gaskets may quickly be cut by placing the gasket material over the flange face. Next tap the edges of the flange with a hammer

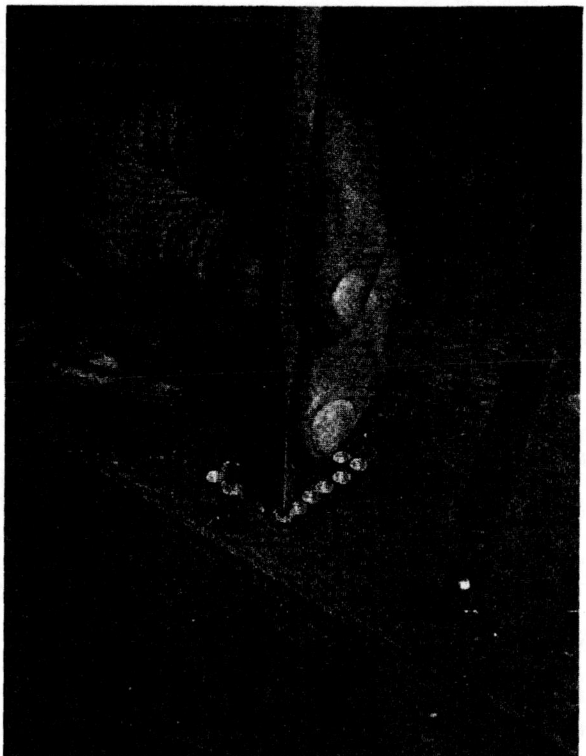

Fig. 19-13 Removing center section of a workpiece

Fig. 19-14 Placing the hacksaw blade in a horizontal position to make long, horizontal cuts

Fig. 19-15 Three hacksaw blades mounted side by side in a hand hacksaw frame

using the flange edge to cut the gasket to size. The flange holes are cut with the ball end of the ball peen hammer. The hacksaw blade may be positioned horizontally in the hand hacksaw to make long, horizontal cuts on a part, Figure 19-14. Two or three saw blades may be placed together on the hand hacksaw blade to widen or make wider screw slots for bolt heads, Figure 19-15.

Repairing a Stripped Internal Thread

A stripped internal thread in a part may be repaired by the following:

1. Select a larger tap drill and drill out the stripped threads.
2. Tap the hole with the correct larger tap.
3. Screw a threaded plug into the hole. This threaded plug can be retained in the hole by welding or using sealant in the threaded joint.
4. Saw off the plug and file the surface flat.
5. Lay out and center punch the correct center position of the hole.
6. Select the correct tap drill size.
7. Drill the hole and countersink the hole.
8. Tap the hole to the correct size, Figure 19-16.

Repairing a Babbitt Bearing

Babbitt metal is an alloy of copper, tin, and antimony. It was used widely for the liners of bearings. Most newer machinery is now made to operate at higher speeds with

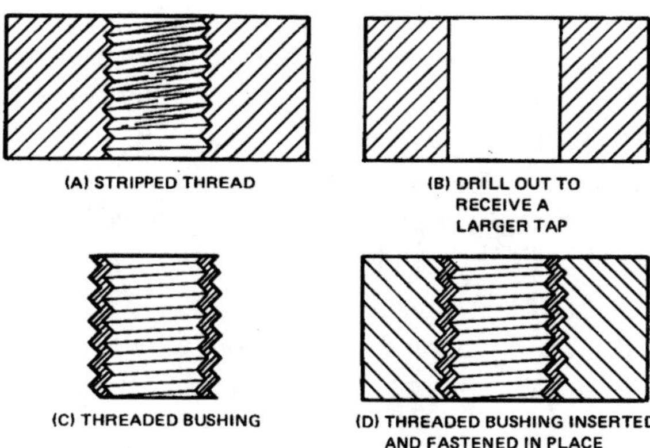

Fig. 19-16 Repairing a stripped internal thread

closer tolerances. The majority of bearings used now are replaceable precision ball or roller bearings.

When babbitting, always make sure eye and face protection are worn.

1. Set up the bearing housing and the shaft as shown in Figure 19-17. They are preheated with a torch to remove moisture and cause the babbit metal to flow freely.

2. The babbitt material is heated slowly in a small melting pot. The temperature is about right when it quickly chars a pine stick to a dark brown, but does not ignite the stick. This temperature is about 870 degrees.

3. Pour the babbitt with a continuous flow. If the pouring is stopped before the entire cavity is filled, the additional metal does not fuse with that already poured. This results in a crack in the bearing. Heat the bottom of the pouring pot and allow the heat to transfer through to melt the babbitt metal.

4. Make sure no water or moisture comes in contact with the melted babbitt. Even a few drops of water causes steam pockets, and the babbitt will spatter violently.

MECHANICAL POWER TRANSMISSION PRODUCTS

Figure 19-18 shows a few of the typical standard mechanical power transmission products available. The products shown in Figure 19-18 are listed below.

The benchworker is required to install, service, and replace these items from time to time. These products are made by various industrial companies. The Dodge Engineering Catalog shown in Figure 19-18 is an excellent source of information in the selection and application of these products.

A. D-78 Engineering Catalog
B. Double-strand chain sprocket with B-style hub
C. Single-strand chain sprocket with B-style hub
D. Single-strand chain

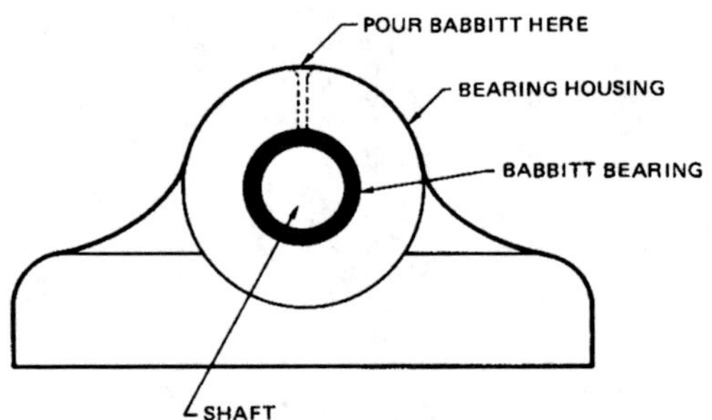

Fig. 19-17 Proper setup of shaft and housing to pour the babbitt bearing

Fig. 19-18 Various mechanical power transmission products

E. Single-strand, chain-plate sprocket – style A
F. Chain spring-clip, connecting-link assembly
G. Chain one-pitch offset link
H. Taper-lock bore, single-strand chain sprocket B-style hub
I. Taper lock bushing
J. Set screws to tighten taper-lock bushing into sprocket
K. Shaft
L. Pillow block bearing
M. Shaft coupling hub
N. Coupling flexible element
O. 2-Bolt flange bearing
P. Bevel gear
Q. 4-Bolt flange bearing
R. Shaft set collar
S. Shaft
T. V belt
U. Taper-lock bushing assembly
V. V-belt sheave
W. Spur gears

REVIEW QUESTIONS

A. Short Answer

1. State the reason why visualization is important to job completion.
2. What are the names of the tools used to restore external threads and internal threads?
3. State three methods to remove a broken tap.
4. Which method of removing a broken tap is probably the most effective?
5. List three methods used to remove a broken stud.
6. Describe how the center section of a workpiece may be removed with a drill and chisel.

ACTIVITIES

1. With your instructor, discuss how to solve several shop bench problems discussed in this unit. Try to think how to get the job done without using standard methods.
2. Remove a broken tap or stud by one of the methods described in this unit.

Table 1   Material Identification Chart

Table 2   Useful Rules

Table 3   Metric to English Conversion

Table 4   English to Metric Conversion

Table 5   RPM Chart for Diameters of Hole Saws

Table 6   Metric Tap Drill Sizes

Table 7   Tap Drill Sizes for Unified Screw Threads with Approximately 75% Depth

Table 8   Decimal Equivalents of Number Size Drills

Table 9   Decimal Equivalents of Letter Size Drills

Table 10   Decimal Equivalents of an Inch by 1/64ths

Table 1 Material Identification Chart

Material	Color	Weight	Spark	Magnetic	Chip	Ring	Filing
Low carbon steel (mild steel)	Dark Gray	Heavy	See Fig. 3-15	Strongly	Continuous chip, smooth edges, chips easily	Distinct ring	Files easily
Medium carbon steel	Dark Gray	Heavy	See Fig. 3-15	Strongly	Continuous chip, smooth edges, chips easily	Distinct ring	Medium
High carbon steel	Dark gray	Heavy	See Fig. 3-15	Strongly	Hard to chip	Distinct ring	Hard to file
Stainless steel	Bright Silvery, Smooth	Heavy	See Fig. 3-15	300 Series No 400 Series Yes	Continuous chip, smooth, bright color	Distinct ring	Medium
Manganese	Dark gray Darker than gray cast iron	Heavy	See Fig. 3-15	Nonmagnetic	Chipped in annealed condition, long chip, difficult	Dull	Difficult in annealed condition
Gray cast iron	Dull gray Evidence of sand mold	Heavy	See Fig. 3-15	Strongly	Small chips, broken 1/8 inch long, not easy, brittle	Dull	Easily filed below surface
Malleable cast iron	Dull gray Evidence of sand mold	Heavy	See Fig. 3-15	Strongly	Continuous chip on exterior, brittle inside	Dull	Files easily
White cast iron	Gray	Heavy	See Fig. 3-15	Strongly	Cannot chip, too hard	Very distinct ring	Cannot be filed, too hard
Copper	Reddish brown or copper	Heavy	None	Nonmagnetic	Continuous chip, smooth edges, soft and easily cut	Dull	Very easily filed, brighter color after filing
Brass	More yellow than copper, more zinc	Heavy	None	Nonmagnetic	Continuous chip, smooth edges, soft and easily cut	Dull	Very easily filed, brighter color after filing
Bronze	More coppery in color than brass	Heavy	None	Nonmagnetic	Continuous chip, smooth edges, soft and easily cut	Dull	Very easily filed, brighter color after filing
Lead	Dark gray	Very Heavy	None	Nonmagnetic	Continuous chip, smooth edges, soft and easily cut	No ring	Easily
Aluminum	White or very light gray	Very light	None	Nonmagnetic	Continuous chip, smooth edges, soft and easily cut	Dull to fairly distinct depending on alloy	Easily
Magnesium	White or very light gray	Very light	None	Nonmagnetic	Small chips, very brittle, easily chipped	Slight	Easily, very bright when filed, filings very fine
White metals	Gray	Heavy	None	Nonmagnetic	Easily cut, continuous chips	Dull	Easily filed like aluminum except heavier

Table 2 Useful Rules

TO FIND CIRCUMFERENCE—
Multiply diameter by 3.1416
Or divide diameter by. 0.3183
TO FIND DIAMETER—
Multiply circumferency by 0.3183
Or divide circumference by 3.1416
TO FIND RADIUS—
Multiply circumference by 0.15915
Or divide circumference by 6.28318
TO FIND SIDE OF AN INSCRIBED SQUARE—
Multiply diameter by 0.7071
Or multiply circumference by 0.2251
Or divide circumference by 4.4428
TO FIND SIDE OF AN EQUAL SQUARE—
Multiply diameter by 0.8862
Or divide diameter by. 1.1284
Or multiply circumference by 0.2821
Or divide circumference by 3.545

SQUARE—
A side multiplied by 1.4142 equals diameter of its circumscribing circle.
A side multiplied by 4.443 equals circumference of its circumscribing circle.
A side multiplied by 1.128 equals diameter of an equal circle.
A side multiplied by 3.547 equals circumference of an equal circle.

TO FIND THE AREA OF A CIRCLE—
Multiply circumference by one-quarter of the diameter.
Or multiply the square of diameter by. 0.7854
Or multiply the square of circumferency by 0.07958
Or multiply the square of 1/2 diameter by 3.1416

TO FIND THE SURFACE OF A SPHERE OR GLOBE—
Multiply the diameter by the circumference.
Or multiply the square of diameter by. 3.1416
Or multiply four times the square of radius by 3.1416

TO FIND THE CUBIC INCHES (VOLUME) IN A SPHERE OR GLOBE—
Multiply the cube of the diameter by 0.5236.

TO FIND THE WEIGHT OF BRASS AND COPPER SHEETS, RODS AND BARS—
Figure the number of cubic inches in piece and multiply same by weight
per cubic inch.

| Aluminum | 0.0924 | Copper | 0.3184 |
| Brass | 0.2960 | Steel | 0.2816 |

Or multiply the length by the breadth (in feet) and product by weight in pounds
per square foot.

Table 3 Metric to English Conversion

Millimetres to Inches					
Milli-metres	Inches	Milli-metres	Inches	Milli-metres	Inches
0.01	0.0004	0.35	0.0138	0.68	0.0268
0.02	0.0008	0.36	0.0142	0.69	0.0272
0.03	0.0012	0.37	0.0146	0.70	0.0276
0.04	0.0016	0.38	0.0150	0.71	0.0280
0.05	0.0020	0.39	0.0154	0.72	0.0283
0.06	0.0024	0.40	0.0157	0.73	0.0287
0.07	0.0028	0.41	0.0161	0.74	0.0291
0.08	0.0031	0.42	0.0165	0.75	0.0295
0.09	0.0035	0.43	0.0169	0.76	0.0299
0.10	0.0039	0.44	0.0173	0.77	0.0303
0.11	0.0043	0.45	0.0177	0.78	0.0307
0.12	0.0047	0.46	0.0181	0.79	0.0311
0.13	0.0051	0.47	0.0185	0.80	0.0315
0.14	0.0055	0.48	0.0189	0.81	0.0319
0.15	0.0059	0.49	0.0193	0.82	0.0323
0.16	0.0063	0.50	0.0197	0.83	0.0327
0.17	0.0067	0.51	0.0201	0.84	0.0331
0.18	0.0071	0.52	0.0205	0.85	0.0335
0.19	0.0075	0.53	0.0209	0.86	0.0339
0.20	0.0079	0.54	0.0213	0.87	0.0343
0.21	0.0083	0.55	0.0217	0.88	0.0346
0.22	0.0087	0.56	0.0220	0.89	0.0350
0.23	0.0091	0.57	0.0224	0.90	0.0354
0.24	0.0094	0.58	0.0228	0.91	0.0358
0.25	0.0098	0.59	0.0232	0.92	0.0362
0.26	0.0102	0.60	0.0236	0.93	0.0366
0.27	0.0106	0.61	0.0240	0.94	0.0370
0.28	0.0110	0.62	0.0244	0.95	0.0374
0.29	0.0114	0.63	0.0248	0.96	0.0378
0.30	0.0118	0.64	0.0252	0.97	0.0382
0.31	0.0122	0.65	0.0256	0.98	0.0386
0.32	0.0126	0.66	0.0260	0.99	0.0390
0.33	0.0130	0.67	0.0264	1.00	0.0394
0.34	0.0134				

Table 4 English to Metric Conversion

Inches to Millimetres					
Inches	Milli-metres	Inches	Milli-metres	Inches	Milli-metres
0.001	0.025	0.290	7.37	0.660	16.76
0.002	0.051	0.300	7.62	0.670	17.02
0.003	0.076	0.310	7.87	0.680	17.27
0.004	0.102	0.320	8.13	0.690	17.53
0.005	0.127	0.330	8.38	0.700	17.78
0.006	0.152	0.340	8.64	0.710	18.03
0.007	0.178	0.350	8.89	0.720	18.29
0.008	0.203	0.360	9.14	0.730	18.54
0.009	0.229	0.370	9.40	0.740	18.80
0.010	0.254	0.380	9.65	0.750	19.05
0.020	0.508	0.390	9.91	0.760	19.30
0.030	0.762	0.400	10.16	0.770	19.56
0.040	1.016	0.410	10.41	0.780	19.81
0.050	1.270	0.420	10.67	0.790	20.07
0.060	1.524	0.430	10.92	0.800	20.32
0.070	1.778	0.440	11.18	0.810	20.57
0.080	2.032	0.450	11.43	0.820	20.83
0.090	2.286	0.460	11.68	0.830	21.08
0.100	2.540	0.470	11.94	0.840	21.34
0.110	2.794	0.480	12.19	0.850	21.59
0.120	3.048	0.490	12.45	0.860	21.84
0.130	3.302	0.500	12.70	0.870	22.10
0.140	3.56	0.510	12.95	0.880	22.35
0.150	3.81	0.520	13.21	0.890	22.61
0.160	4.06	0.530	13.46	0.900	22.86
0.170	4.32	0.540	13.72	0.910	23.11
0.180	4.57	0.550	13.97	0.920	23.37
0.190	4.83	0.560	14.22	0.930	23.62
0.200	5.08	0.570	14.48	0.940	23.88
0.210	5.33	0.580	14.73	0.950	24.13
0.220	5.59	0.590	14.99	0.960	24.38
0.230	5.84	0.600	15.24	0.970	24.64
0.240	6.10	0.610	15.49	0.980	24.89
0.250	6.35	0.620	15.75	0.990	25.15
0.260	6.60	0.630	16.00	1.000	25.40
0.270	6.86	0.640	16.26		
0.280	7.11	0.650	16.51		

Table 5 RPM Chart for Diameters of Hole Saws

Saw Diameter		Material				
Inches	Milli-metres	Mild Steel	Tool and Stainless Steel	Cast Iron	Brass	Aluminum
5/8	16	550	275	365	730	825
1	25	350	160	215	435	480
1 1/2	38	230	115	150	300	345
2	51	170	85	115	230	255
3	76	115	55	75	150	170
4	102	85	40	55	110	130
6	152	55	25	35	75	85

Table 6 Metric Tap Drill Sizes

Metric Tap Size	RECOMMENDED METRIC DRILL				CLOSEST RECOMMENDED INCH DRILL			
	Drill Size mm	Inch Equiv.	Probable Hole Size (Inches)	Probable Percent of Thread	Drill Size	Inch Equiv.	Probable Hole Size (Inches)	Probable Percent of Thread
M1.6 X 0.35	1.25	0.0492	0.0507	69	—	—	—	—
M1.8 X 0.35	1.45	0.0571	0.0586	69	—	—	—	—
M2 X 0.4	1.60	0.0630	0.0647	69	#52	0.0635	0.0652	66
M2.2 X 0.45	1.75	0.0689	0.0706	70	—	—	—	—
M2.5 X 0.45	2.05	0.0807	0.0826	69	#46	0.0810	0.0829	67
M3 X 0.5	2.50	0.0984	0.1007	68	#40	0.0980	0.1003	70
M3.5 X 0.6	2.90	0.1142	0.1168	68	#33	0.1130	0.1156	72
M4 X 0.7	3.30	0.1299	0.1328	69	#30	0.1285	0.1314	73
M4.5 X 0.75	3.70	0.1457	0.1489	74	#26	0.1470	0.1502	70
M5 X 0.8	4.20	0.1654	0.1686	69	#19	0.1660	0.1692	68
M6 X 1	5.00	0.1968	0.2006	70	#9	0.1960	0.1998	71
M7 X 1	6.00	0.2362	0.2400	70	15/64	0.2344	0.2382	73
M8 X 1.25	6.70	0.2638	0.2679	74	17/64	0.2656	0.2697	71
M8 X 1	7.00	0.2756	0.2797	69	J	0.2770	0.2811	66
M10 X 1.5	8.50	0.3346	0.3390	71	Q	0.3320	0.3364	75
M10 X 1.25	8.70	0.3425	0.3471	73	11/32	0.3438	0.3483	71
M12 X 1.75	10.20	0.4016	0.4063	74	Y	0.4040	0.4087	71
M12 X 1.25	10.80	0.4252	0.4299	67	27/64	0.4219	0.4266	72
M14 X 2	12.00	0.4724	0.4772	72	15/32	0.4688	0.4736	76
M14 X 1.5	12.50	0.4921	0.4969	71	—	—	—	—
M16 X 2	14.00	0.5512	0.5561	72	35/64	0.5469	0.5518	76
M16 X 1.5	14.50	0.5709	0.5758	71	—	—	—	—
M18 X 2.5	15.50	0.6102	0.6152	73	39/64	0.6094	0.6144	74
M18 X 1.5	16.50	0.6496	0.6546	70	—	—	—	—
M20 X 2.5	17.50	0.6890	0.6942	73	11/16	0.6875	0.6925	74
M20 X 1.5	18.50	0.7283	0.7335	70	—	—	—	—
M22 X 2.5	19.50	0.7677	0.7729	73	49/64	0.7656	0.7708	75
M22 X 1.5	20.50	0.8071	0.8123	70	—	—	—	—
M24 X 3	21.00	0.8268	0.8327	73	53/64	0.8281	0.8340	72
M24 X 2	22.00	0.8661	0.8720	71	—	—	—	—
M27 X 3	24.00	0.9449	0.9511	73	15/16	0.9375	0.9435	78
M27 X 2	25.00	0.9843	0.9913	70	63/64	0.9844	0.9914	70
M30 X 3.5	26.50	1.0433						
M30 X 2	28.00	1.1024						
M33 X 3.5	29.50	1.1614						
M33 X 2	31.00	1.2205						
M36 X 4	32.00	1.2598		REAMING RECOMMENDED TO THE DRILL SIZE SHOWN				
M36 X 3	33.00	1.2992						
M39 X 4	35.00	1.3780						
M39 X 3	36.00	1.4173						

FORMULA FOR METRIC TAP DRILL SIZE

$$\text{Basic Major Dia. (mm)} - \frac{\% \text{ Thread} \times \text{Pitch (mm)}}{76.980} = \text{DRILLED HOLE SIZE (mm)}$$

FORMULA FOR PERCENT OF THREAD

$$\frac{76.980}{\text{Pitch (mm)}} \times \left[\text{Basic Major Dia. (mm)} - \text{Drilled Hole Size (mm)} \right] = \text{Percent of Thread}$$

Table 7 Tap Drill Sizes for Unified Screw Threads with Approximately 75% Depth

Unified Coarse			Unified Fine		
Tap Size	Threads per inch	Tap Drill Size	Tap Size	Threads per inch	Tap Drill Size
# 5	40	# 38	# 5	44	# 37
# 6	32	# 36	# 6	40	# 33
# 8	32	# 29	# 8	36	# 29
# 10	24	# 25	# 10	32	# 21
# 12	24	# 16	# 12	28	# 14
1/4	20	# 7	1/4	28	# 3
5/16	18	F	5/16	24	I
3/8	16	5/16	3/8	24	Q
7/16	14	U	7/16	20	25/64
1/2	13	27/64	1/2	20	29/64
9/16	12	31/64	9/16	18	33/64
5/8	11	17/32	5/8	18	37/64
3/4	10	21/32	3/4	16	11/16
7/8	9	49/64	7/8	14	13/16
1	8	7/8	1	14	15/16
1-1/8	7	63/64	1-1/8	12	1-3/64
1-1/4	7	1-7/64	1-1/4	12	1-11/64
1-3/8	6	1-7/32	1-3/8	12	1-19/64
1-1/2	6	1-11/32	1-1/2	12	1-27/64
1-3/4	5	1-9/16			
2	4-1/2	1-25/32			

Table 8 Decimal Equivalents of Number Size Drills

No.	Size of Drill in Inches	No.	Size of Drill in Inches	No.	Size of Drill in Inches	No.	Size of Drill in Inches
1	0.2280	21	0.1590	41	0.0960	61	0.0390
2	0.2210	22	0.1570	42	0.0935	62	0.0380
3	0.2130	23	0.1540	43	0.0890	63	0.0370
4	0.2090	24	0.1520	44	0.0860	64	0.0360
5	0.2055	25	0.1495	45	0.0820	65	0.0350
6	0.2040	26	0.1470	46	0.0810	66	0.0330
7	0.2010	27	0.1440	47	0.0785	67	0.0320
8	0.1990	28	0.1405	48	0.0760	68	0.0310
9	0.1960	29	0.1360	49	0.0730	69	0.0292
10	0.1935	30	0.1285	50	0.0700	70	0.0280
11	0.1910	31	0.1200	51	0.0670	71	0.0260
12	0.1890	32	0.1160	52	0.0635	72	0.0250
13	0.1850	33	0.1130	53	0.0595	73	0.0240
14	0.1820	34	0.1110	54	0.0550	74	0.0225
15	0.1800	35	0.1100	55	0.0520	75	0.0210
16	0.1770	36	0.1065	56	0.0465	76	0.0200
17	0.1730	37	0.1040	57	0.0430	77	0.0180
18	0.1695	38	0.1015	58	0.0420	78	0.0160
19	0.1660	39	0.0995	59	0.0410	79	0.0145
20	0.1610	40	0.0980	60	0.0400	80	0.0135

Table 9 Decimal Equivalents of Letter Size Drills

Letter	Inches	Letter	Inches	Letter	Inches
A	0.234	J	0.277	S	0.348
B	0.238	K	0.281	T	0.358
C	0.242	L	0.290	U	0.368
D	0.246	M	0.295	V	0.377
E	0.250	N	0.302	W	0.386
F	0.257	O	0.306	X	0.397
G	0.261	P	0.323	Y	0.404
H	0.266	Q	0.332	Z	0.413
I	0.272	R	0.339		

Table 10 Decimal Equivalents of an Inch by 1/64ths

		1/64	0.015625	33/64	0.515625
	1/32		0.03125	17/32	0.53125
		3/64	0.046875	35/64	0.546875
1/16			0.0625	9/16	0.5625
		5/64	0.078125	37/64	0.578125
	3/32		0.09375	19/32	0.59375
		7/64	0.109375	39/64	0.609375
1/8			0.125	5/8	0.625
		9/64	0.140625	41/64	0.640625
	5/32		0.15625	21/32	0.65625
		11/64	0.171875	43/64	0.671875
3/16			0.1875	11/16	0.6875
		13/64	0.203125	45/64	0.703125
	7/32		0.21875	23/32	0.71875
		15/64	0.234375	47/64	0.734375
1/4			0.25	3/4	0.75
		17/64	0.265625	49/64	0.765625
	9/32		0.28125	25/32	0.78125
		19/64	0.296875	51/64	0.796875
5/16			0.3125	13/16	0.8125
		21/64	0.328125	53/64	0.828125
	11/32		0.34375	27/32	0.84375
		23/64	0.359375	55/64	0.859375
3/8			0.375	7/8	0.875
		25/64	0.390625	57/64	0.890625
	13/32		0.40625	29/32	0.90625
		27/64	0.421875	59/64	0.921875
7/16			0.4375	15/16	0.9375
		29/64	0.453125	61/64	0.953125
	15/32		0.46875	31/32	0.96875
		31/64	0.484375	63/64	0.984375
1/2			0.5	1	1.

ACKNOWLEDGMENTS

Illustrations

The author wishes to thank the following companies for their contributions of illustrations and information.

Adjustable Clamp Company
American National Standards Institute
American Society of Mechanical Engineers
Baldor Electric Company
Behr Manning, Division of Norton Company
Bondhus Tool Company, Inc.
The Challenge Machinery Company
Channellock, Inc.
Chicago Dial Indicator Company
Cleveland Twist Drill, an Acme-Cleveland Company
Dake Corporation
DoAll Company
The Du Mont Corporation
Dykem Company
ESNA Division of Amerace Corporation
Greenfield Tap & Die, Division of TRW, Inc.
Groov-Pin Corporation
Jaw Manufacturing Company
Marson Fastener Corporation
Milwaukee Electric Tool Corporation
MTI Corporation
Nicholson File Company
Norton Company, Norton Abrasive Materials Division
OTC, Division of Owatonna Tool Company
Precision Brand Products
Proto Professional Tools, Division of Ingersoll-Rand Company
The Ridge Tool Company

Simonds Cutting Tools Division, Division of Wallace Murray Corporation
Sioux Tools, Inc.
Skil Corporation
Standard Steel Company
The Stanley Works
L. S. Starrett Company
Swiss Precision Instruments, Inc.
Tapmatic Corporation
Titan Tool Supply Company
Waldes Kohinoor, Inc.
Walton Company
Wilton Corporation

The author wishes to thank Linda Handy, Debbie Scarbrough, Tran Van Nhung, Shirley Lamkey, and Everett Koontz for their contributions to the text. The author expresses thanks to those persons who helped in many ways in the practical completion of the text. Special appreciation is given to Bob Shaw for his outstanding professional photography and dedicated assistance.

Reviewer

David L. Taylor

Delmar Staff

Mark W. Huth, Sponsoring Editor
Barbara A. Christie, Associate Editor

Classroom Testing

The materials and information included in this text have been classroom tested at Chemeketa Community College, Salem, Oregon.

Index

A

Abrasive cloth, 126–127
Abrasive cutoff saw, 184
Abrasive stick dressers, grinding wheel, 182
Abrasive type grinders, 180
Accuracy, layout, 99
    measurement, 42
Acknowledgments, 202–203
Acme taps, 150
Adhesive fastening, 34
Adjustable T-handle tap wrench, 148
Adjustable wrench, 109
Air-powered tools, 159
AISI, steel numbering system, 17
Aligning fasteners, 30–31
Aligning punch, 114
Allowance, defined, 71
Allowance measurement systems, 73
Allowances for fits, 74
   chart, 75
Alloy steels, 12
Alloys, copper-base, 13
Aluminum, 13
Aluminum oxide, grinding, 180
American National Threads, 147
Anaerobic adhesives, 34
Angle plate, 92
Annealing, 174
Arbor press, 114–115
Attitudes, 1

B

Babbitt bearing, repairing, 191
Ball peen hammers, 109
Band saw blades, 166–168
Bar stock, 7
Bars and punches, 113–114
Basic hole system, 73
Basic shaft system, 73

Bastard file, 124
Beam trammel, 84
Bench and pedestal grinders, 178–184
Bench block, 88
Bench plate, 88, 89
Bench processes, miscellaneous, 190–192
Bench tools, 3
Bench vise, 105–107
Benchwork, basics, 1–4
Benchworker, duties, 1
Beryllium copper, 13–14
    hazard, 14
Bevel protractor, combination set, 86, 87
Bilateral tolerance, 71
Bill of materials, 20
Blade installation, horizontal band saw, 171
    vertical band saw, 172
Blades, band saw, 166–168
Blind rivet, 30
Blueprints, care, 98
Bolts, 26
    grading, 25
Bond type grinding wheel, 180
Bondus-style hex wrench, 110, 111
Bottoming tap, 146, 147
Box wrenches, 110
Brass, 13
Brazing, 34
Broaches, 136–137
Broken stud, removal, 190
Broken tap, removing, 187–190
Bronze, 13
Butt welder, 173

C

C clamps, 107
Calipers, 47–51
   hermaphrodite, 84–86
   Metric, 58–59, 64
   micrometer, 55–62
   slide, 65
   Vernier, 57–65

Cap and acorn nuts, 28
Cap screws, 27
    removal, 190
Cape chisel, 118
Carbon, 14
Carbon steels, 11
    numbering system, 17
Carriage clamps, 107
Case-hardened steel, 12
Cast iron, 10
Castellated and slotted nuts, 28
Casting layout, 99
Castings, 10
Caution, *see* Safety precautions
Center drill, 130
Center head, combination set, 86, 87
Center punch, 80–81, 114, 183
Center section of workpiece, removing, 190
Centers, locating, 86
    machined parts, 100
Ceramic, 14
Chain wrench, 108, 111
Chamfering, defined, 129
Chipping test, 19
Chisels, 117–121
Chords, table of, 100, 102
Chromium, alloying metal, 13
Circle, finding area of, 196
    using dividers, 83
Circumference, finding, 196
Clamps, 81, 107
Clearance fit, 72
Coarse threads, 24
Cobalt, alloying metal, 13
Cold chisel, 118
Cold-forming taps, 150
Cold-rolled mild steel, 12
Cold saw, 184
Color-code marking system, steel, 17
Combination set, 86–87
Combined drill and countersink, 130
Compression stop nuts, 28
Contact measurements, 44
Conversion, English and Metric systems, 39
    tables, 197
Copper, 13
Copper-base alloys, 13
Copper-base materials, 14
Corrosion, 7
Cotter pins, 30
Counterboring, defined, 129
Countersinking, defined, 129
Countersinks, 126
Cross-peen hammer, 109
Curved cut file, 122
Cut of files, 121
Cutting fluids, 148–149, 169

D

Deburring, 126
Decimal equivalents, of inch, 38, 201
    millimetre, 39
    number and letter size drills, tables, 200

Decimal rules, 46
Depth rule gage, 45
Diagonal cutters, pliers, 108
Dial comparator, 66
Dial indicators, 65–66
Dial snap gages, 66
Diameter, finding, 196
Diamond dressers, grinders, 182
Diamond-point chisel, 118
Diamond-pointed wheel dressers, 17
Die, cutting thread with, 154–155
Die grinder, 163
Die stock, 153
Direct measurements, 43
Dividers, 82–83
Double cut file, 122, 124
Double-ended adjustable tap wrench, 148
Double square, 88
Dowel pins, 30
Drawfiling, 123
Dressing grinding wheel, 181–182
Drift punch, 114
Drill chuck, 160
Drill grinding procedure, 131–133
Drill point gages, 51
Drill presswork, layout, 98
Drill-size gages, 51
Drills, 129–134
    decimal equivalents of, table, 200
Drive screws, 27

E

Elastic stop nuts, 28
Electric-powered tools, 159–193
Emery cloth, 126
End-milled keyways, 33
English measurement system, 38
    conversion, 39
    conversion table, 197
Expanding rivets, 30
Expansion fit, 73
Extension taps, 150
External threads, 153
Extruded shapes, 14
Eye protection, *see* Goggles

F

Fasteners, 23–34
    adhesive, 34
    brazing, 34
    grading for strength, 25
    identification, 23
    keys, 32–34
    nonthreaded, 29–31
    retaining, 31–32
    rule, 26
    sizes and selections, 24
    soldering, 34
    storage, 34
    threaded, 26–29
    unified threads, 24
    welding, 34

Fastening rule, 26
Feathered key, 33
Feeler gages, 76
Ferrous metals, 10
File test, 19
Files, 121–125
Fine threads, 24
Finished surfaces, layout, 100
Firm-joint caliper, 48
Fit, 72–73
Fitting instructions, 70
Fitting terminology, 70
Fittings, allowances for, 74
   chart of allowances, 75
   operations, 76
   print specifications, 74
   tolerances for, 74
Fixed fasteners, 29
Flat chisel, 118
Flat files, 122
Flat filing, 123
Flat scrapers, 125
Flat washers, 29
Flexible extension gripping tool, 187
Floor work, 1
Flowering process, 126
Flutes, 145
Forced fit, 73
Forgings, 9
Fractional divisions of inch, 38, 201
   of millimetre, 39
Frosting process, 126

G

Gage blocks, 67
Gages, 51–52
Gasket, 190
Gib head taper keys, 33
Globe, finding surface and volume, 196
Goggles, needed for bars and punches, 114
   broaching, 137
   rotary files, 141
   tap removal, 188, 189
   tapping operations, 151, 152
   using chisels, 119
   See also Safety precautions
Grading fasteners, 25
Gray cast iron, 10
Grinders, 162
   bench and pedestal, 178–184
Grinding wheel, 178
   dressing, 181
   mounting, 181
Ground burs, 141
Group layout, 98

H

Hacksaw, hand, 137–140
Half-round files, 122
Half-round scrapers, 125
Hammers, 108–109
   layout, 80

Hand grinder, 162
Hand hacksaw, 137–140
Hand reamer, 134
Hand saws, 137–141
Hand tap, 144–147
Hand tools, 1, 105–147
Hardness test, 19
Heat treating, 12
Heavy nuts, 28
Hermaphrodite caliper, 84–86
Hex nut, 27
Hex socket head wrenches, 110
Hexagon head bolts, 26
Hexagonal, rethreading die, 153
High-carbon steel, 12
High shear taps, 150
High-speed steel, 12
High-temperature metals, 14
Hole saws, 140–141
   RPM chart for diameters of, 197
Hole system, basic, 73
Hook rule, 44, 45
Hook scrapers, 125
Horizontal band saw, 169–171

I

Identification chart, materials, 195
Identifying steels, 16–19
Inch, decimal equivalents of, 38
   table, 201
Indirect measurements, 43
Industrial fasteners, *see* Fasteners
Industrial metals, types, 10
   identifying materials, 195
Insert die, 153
Inside micrometers, 62–63
Interference fit, 72, 73
Internal thread, repairing, 191
Internal threaded fasteners, 27
Internal threaded inserts, 29
Interrupted thread taps, 150
Iron, 10
Iron-carbon spark test, 17

J

Jam nut, 28
Jigsaw, 162
Johnson key, 33

K

Keys, 32–34
Keyseat, 32
Keyway, 32–33
   cutting, 136

L

Layout fluid, 98
Layout hammer, 80
Layout procedures, 97–103
Layout tools, nonprecision, 79–82
   precision, 91–96
   semiprecision, 82–89

Left-hand threads, 25
Letter size drills, decimal equivalents, 200
Levels, precision, 67
Light waves, 37
Linear measurement, 38–39
Lock-joint caliper, 48
Lock washers, 29
Locking pliers, 108
Low-carbon hot-rolled steel, 11

M

Machinability, 12
Machine bolts, 26
Machine screws, 27
Machine steel, 11
Machine work, defined, 1
Machined parts, layout, 100
Machinist's vise, 105–107
Magnesium, 13
Magnet test, 19
Malleable iron, 11
Manganese, alloying metal, 13
Materials, 7–14
    bill of, 20
    high-temperature metals, 14
    identification chart, 195
    industrial metals, 10–14
    metals, basic forms, 7
    nonmetallics, 14
    precious metals, 14
    selection and cuttings, 19–21
    special, 10
    storage, 21
Mathematical rules, 196
    conversion tables, 197
Measure, units of, 37–39
    conversion tables, 197
Measurement, allowance, 73
    conversion tables, 197
    methods, 42–43
Measuring tools, semiprecision, 42–52
    calipers, 47–51
    classification, 44
    gages, 51
    miscellaneous, 51
    rules and tapes, 44–47
    selection and care, 42
Mechanical dresser, grinding wheel, 182
Mechanical fasteners, *see* Fasteners
Mechanical power transmission products, 192
Medium-carbon steel, 12
Metal bolt cutters, 142
Metal-cutting band saw, 166
Metal-cutting snips, 142
Metals, basic forms, 7
    ferrous, 10–13
    high-temperature, 14
    industrial, 10
    nonferrous, 13
    rare and precious, 14
    special, 10

Metric measurement system, 38
    conversion tables, 39, 197
Metric micrometer, 58–59
Metric rules, 46
Metric tap drill sizes, table, 198
Metric threads, 147–148
Metric vernier calipers, 64–65
Micrometer, calipers, 51, 55
    care, 60
    measurement transfer, 61
    measuring with, 59
    Metric, 58–59
    reading, 56
    special purposes, 62
    Vernier, 57–58
Micrometer depth gages, 62
Mild steel, 11
Mill file, 122
Millimetre, decimal equivalents, 39
Mistakes, effect on profits, 2
Molybdenum, 13, 14
Morphy, 84–86
Mounting grinding wheel, 181
Multiple same parts layout, 102–103

N

Needle-nose pliers, 108
Nickel, alloying metal, 13
Nickel-base metals, 14
Nominal diameter size, 12
Nonferrous metals, 13–14
Nonmetallics, 14
Nonprecision layout tools, 79–82
Nonthreaded fasteners, 29–31
Number size drills, decimal equivalents, 200
Numbering systems, identifying steel, 16
Nut, removal, 190
Nut driver, 112, 113
Nuts, 27–28
Nylon spot nut, 28

O

Open-center parts, finding center, 100
Open-end wrenches, 110
Operations used in benchwork, 3

P

Parallel clamps, 81, 107
Parallels, steel, 92
Phillips screwdriver, 112
Pin punches, 113
Pinning, files, 124
Pipe, 7
    dimensions, 9
Pipe taps, 150
Pipe threads, cutting, 155–156
Pipe wrench, 111
Pitch, band saw blade, 168
    defined, 144
Plastics, 14
Pliers, 108

Plug tap, 146, 147
Pocket keyway, 33
Pocket slide caliper rule, 45
Polishing cloths, 126–127
Pop rivet, 31
Power, mechanical power transmission products, 192
Power hand drills, 159–164
Power saws, 166–175
Power tools, 159–193
    See also Tools
Precious metals, 14
Precision layout tools, 91–96
Precision levels, 67
Precision measurements, 37
Precision measuring tools and gages, 54–67
    defined, 44
Precision straightedges, 93
Precision V blocks, 93
Prick punch, 80–81, 114, 183
Print specifications, tolerances, 74
Problem-solving, 186
Procedures, common, 186
Products, mechanical power transmission, 192
Profitable benchwork, guide to, 2
Protractor, steel, 88
    Universal bevel, 94
Pry bar, 113
Pullers, 114
Pulley taps, 149
Punches and bars, 113–114

R

Radius, finding, 196
Radius gages, 52
Rare metals, 14
Rasp cut file, 122
Ratchet stock wrenches, 155
Ratchet wrench, 111
Reamers, 134–136
Reciprocating power hacksaw, 166
Red hardness, 13
Reed and Prince screwdriver, 112
Regular nuts, 28
Repair work, 4
    babbitt bearing, 191
    stripped internal thread, 191
Responsibility to know tools, 54
Retaining fasteners, 31–32
Retaining rings, 31
Rethreading dies, 187
Rethreading taps, 187
Reversing power drills, 161
Right-hand threads, 25
Ring test, 19
Rivet mandrel, 30, 31
Rivets, 29–30
Roll pins, 31
Rotary power files, 141
Round-corner filing, 123
Round-nose chisel, 118
Round-nose pliers, 108

Round solid, die, 153
Round-split adjustable, die, 153
Rubbers, 14
Rules, mathematical, 196

S

SAE steel numbering system, 16
Safe work habits, 1
Safety glasses, *see* Goggles
Safety precautions, bars and punches, 114
    beryllium copper, 14
    fitting operations, 76–77
    grinding, 163–164
    grinding wheel, 183
    injury prevention, 21
    layout tool care, 79
    Metric and English tools, 105
    power drilling, 160, 161
    power sawing, 175
    screwdrivers, 113
    tightening fastener, 26
    using chisels, 119, 121
    using files, 124
    weld flash, 174
    See also Goggles
Scales, 44
Scaling, 103
Scrapers, 125–126
Scratch test, 19
Screw pitch gages, 52
Screw threads, tap drill sizes, table, 199
Screwdriver blade, grinders, 183
Screwdrivers, 112–113
Screws, 26
Scriber, 79–80
Second-cut file, 125
Second tap, 147
Semiprecision layout tools, 82–89
Semiprecision measuring tools, 42–52
    defined, 44
Serial taps, 150
Set, band saw blades, 168
Set screws, 27
Shaft system, basic, 73
Sharpening divider points, 83
Shop press, 115
Shop tests, identifying steel, 17–19
Shrink fit, 73
Side cutters, pliers, 108
Silicon carbide grinding, 180
Sine bar, 67
Single cut file, 121
Single-cut mill file, 125
Size, fitting terminology, 70
Slide calipers, 65
Sliding fit, 72
Slip joint pliers, 108
Slot, cutting, 136
Snap ring pliers, 108
Socket wrenches, 110
Soft-faced hammers, 109

Soldering, 34
Soldering irons, 187
Spanner wrench, 112
Spark test, metals, 17
Special materials, 10
Special spiral-fluted taps, 150
Special thread series, 24
Special tools and procedures, 186–193
Sphere, finding surface and volume, 196
Spiral-fluted reamer, 134
Spirit level, combination set, 86
Split dies, 153
Split dowel pins, 31
Spotfacing process, 130
Spotting process, 125
Spring bow-style calipers, 48
Spring pins, 31
Square, finding sides of, 196
     steel, 92
Square files, 122
Square head, combination set, 86
Square head bolts, 26
Square keys, 32–33
Square nut, 27
Square nut die, 153
Stainless steels, 13
Starter tap, 146
Steel, 11
     methods of identifying, 16–19
Steel parallels, 92
Steel protractor, 88
Steel rule, 44–47
     combination set, 86
Steel square, 92
Steel tapes, 46
Stock, 153, 154
Storage of materials, 21, 34
Straight peen hammer, 109
Straightedges, 87
     precision, 93
Straightforward filing, 123
Strap wrench, 111
Stripped internal thread repairing, 191
Stripping, 26
Stud bolt, 28
Studs, 28
     removing, 190
Surface coatings, 149
Surface gage, 88, 89
Surface grinder, 183
Surface plate, 91
Swiss pattern files, 122

T

Tantalum, 14
Tap, removing, 187–190
Tap cutting, surface coatings, 149
Tap drill sizes, 146–147
     Metric, 147, 198
     unified screw threads, table, 199
Tap extractor, 188

Tap wrenches, 148
Taper keys, 33
Taper pin reamers, 134
Taper pins, 30
Taper tap, 146, 147
Tapping fluids, 149
Tapping threads in workpieces, 151–153
Taps, 144–147
     hand tapping threads, 151–153
     special, 149–150
Thickness gages, 76
Thread cutting with die, 154–155
Thread-forming screws, 27
Threaded fasteners, identification, 23
     kinds, 26–29
     unified threads, 24
Threading dies, 153–155
Threads, defined, 144
     hand tapping in workpiece, 151–153
     Metric, 147–148
Three-cornered scrapers, 125
Three square files, 122
Thumb screws, 27
Titanium, 13
Tolerance, 70–71
     for fitting, 74
Tongue and groove pliers, 108
Tool steel, 12
Toolmaker's buttons, 95–96
Tools, 3
     air-powered, 159
     care, 3, 79
     cutting, 117–127
     electric-powered, 159
     hand, 105–147
     measuring, 42–52
     noncutting, 105–115
     nonprecision, 79–82
     power, 159–193
     precision, 54–67, 91–96
     semiprecision, 82–89
     special, 186–193
Tooth blades, band saw, 166
Torque wrench, 111
Transfer calipers, 49
Transfer punches, 82, 114
Transition fit, 72
Tubing, 7
Tungsten, 14
Twist drills, 130–131

U

Unified threads, 24
Unilateral tolerance, 71
Universal bevel protractor, 94–95

V

V blocks, 88, 89
     precision, 93
Vanadium, alloying metal, 13
Vernier calipers, 63–64
     Metric, 64–65

Vernier height gage, 93
Vernier micrometer, 57–58
Vertical band saw, 171–175
Vises, 105–107
Visualization, 186

W

Washers, 29
Wavelength of light, 37
Weight, brass and copper sheets, 196
Welding, 34
White cast iron, 10

Wing nut, 28
Wing screws, 27
Wireless gages, 51
Witness marks, 80
Wood, 14
Woodruff keys, 32
Work habits, 1
Work plan, 4
    layout, 99
Wrenches, 109–112
    tap, 148
Wrought iron, 11